ROYAL
OBSERVATORY
GREENWICH

T0345794

The Search for Life

Hannah Banyard

Royal Observatory Greenwich
Illuminates

First published in 2024 by Royal Museums Greenwich, Park Row, Greenwich, London, SE10 9NF

ISBN: 978-1-906367-90-9

Text © National Maritime Museum, Greenwich, London

Hannah Banyard has asserted her right under the Copyright, Designs and Patent Act 1988 to be identified as the author of this work.

At the heart of the UNESCO World Heritage Site of Maritime Greenwich are the four world-class attractions of Royal Museums Greenwich – the National Maritime Museum, the Royal Observatory, the Queen's House and *Cutty Sark*.

rmg.co.uk

A CIP catalogue record for this book is available from the British Library.

Typesetting by Thomas Bohm, User Design, Illustration and Typesetting
Cover design by Ocky Murray
Diagrams by Dave Saunders
Printed and bound by Bell & Bain Ltd.

FSC
www.fsc.org
MIX
Paper | Supporting responsible forestry
FSC® C007785

About the Author

Hannah Banyard is a science communicator with a passion for making complex scientific ideas accessible. She was formerly an astronomer at Royal Observatory Greenwich. Following a BSc in Astrophysics, she conducted astrobiology research during an MSc in Planetary Science. She believes the visibility of women in STEM roles is key to encouraging more diversity in the scientific community and regularly gives expert media appearances and talks.

About Royal Observatory Greenwich

The historic Royal Observatory has stood atop Greenwich Hill since 1675 and documents over 800 years of astronomical observation and timekeeping. It is truly the home of space and time, with the world-famous Greenwich Meridian Line, awe-inspiring astronomy and the Peter Harrison Planetarium. The Royal Observatory is the perfect place to explore the Universe with the help of our very own team of astronomers. Find out more about the site, book a planetarium show, or join one of our workshops or courses online at rmg.co.uk.

Contents

In loving memory of Cameron,
Whose eyes I opened to the stars,
Who, in turn, opened mine to the Earth.

Introduction

Questions about whether humanity is alone in the Universe are not new. The modern world has widened our view of what *could* be, through advances in technology and space exploration, but it is not responsible for the first postulations of life existing elsewhere. It seems intrinsically human to query our place within the cosmos, perhaps induced by observing the vast sea of stars in the night skies above us or perhaps without any prompting at all – like a sudden intrusive thought. As early as 600 BC, the Greek philosopher Anaximander championed the idea of **cosmic pluralism**, the existence of multiple worlds other than Earth that harbour extraterrestrial life. Some 200 years later, Plato and Aristotle, the 'great thinkers', opposed this idea and advocated instead a unique Earth, dismissing even the very thought of life elsewhere and suppressing progression on the topic for nearly a millennium. By the medieval era, though, some Islamic scholars, among them Ibn Abbas (*c*.619–*c*.687 CE), supported the idea of extraterrestrial life.

Throughout this book we will traverse the very human story of life on Earth and how scientists are actively seeking signs of life beyond it. We will delve into the origins of life, what constitutes life and how we are looking to our own Solar System and beyond for answers. The search for life is not just a pursuit of the unknown; it is fundamental to further understanding ourselves and where we come from, and it may well dictate our future.

Some find the question of determining what life is so profound that they spend their lives attempting to answer it.

The study of life in the Universe is a relatively new field of science known as **astrobiology**, an interdisciplinary subject that focuses on the origins of life, its evolution and distribution in the Universe. Prompted by the dawn of space exploration, Gavriil Tikhov (1875–1960) first proposed the term in 1953. Tikhov was a Soviet astronomer and a pioneer in a field that requires a comprehensive understanding of biology, cosmology and planetary science. A lack of certifiable knowledge is what makes the subject so tantalising. Just 120 years ago, it was thought that everything there was to know about physics could be described by the theories and principles that had already been established (what we now term 'classical physics') and that any further research would involve fine-tuning these concepts. But the discovery of x-rays and sub-atomic particles in the early 20th century cast doubt on these conclusions, heralding a new era of modern physics. Science is never complete: every time something new is learned, more questions arise.

You might be thinking, 'Hold on, didn't we find life elsewhere some years ago?' The short answer is no. We have never found any definitive evidence of alien life. We have, however, had controversial evidence, which has led to considerable misconceptions and quite understandably so. In 1984, a meteorite (ALH84001) was picked up in Allan Hills, Antarctica and studied by NASA scientists. Around ten years later they determined that it was a Martian meteorite, a piece of rock from the surface of Mars which had been knocked off the planet after a collision some 17 million years ago. The rock then hurtled through space towards Earth, settling on our planet around 13,000 years before being collected. Compositionally, ALH84001 was found to have traces of gas matching the atmosphere of Mars,[1]

---

[1] Over 95% carbon dioxide, with trace amounts of nitrogen, argon, oxygen and other gases.

which identified it as an SNC meteorite (the group name given to meteorites that originated on Mars). It is an intriguing piece of rock that gives us an insight into the Martian environment, and it's all the more exciting for the fact that radiometric dating (a calculation of the age of a material made by measuring the presence of radioactive elements) told us it was more than 4 billion years old. ALH84001 is, therefore, not only much older than the other SNC meteorites found on Earth, but is also believed to have been part of Mars at an earlier time, when the planet was covered in water. This finding made the meteorite unique and thus of great interest – the importance of water when it comes to potential extraterrestrial life cannot be understated. On Earth, at least, there is no life without water.

In 1996, having continued their studies, NASA scientists believed they had discovered a sign of life in ALH84001. The headline 'Life Found on Mars' was plastered all over the front pages of newspapers around the world with a picture of the meteorite, which contained what looked to the untrained eye like a tiny worm. (Even the then President of the United States, Bill Clinton, made a televised announcement to celebrate the findings.) At no point did the scientists themselves believe this to be a fossilised worm, but they did think it could perhaps be a tiny microbe – an organism too small to be visible to the naked eye. It was essentially a microscopic chain of molecules (atoms bonded together) approximately 20–500 nanometres long and it was proposed the organism could be a nanobacterium, a theorised type of life smaller than any cellular life known on Earth. There were supposedly three different properties possessed by ALH84001 that indicated life: the direct imaging of the 'microbe' (image 1), the presence of other organic material (pertaining to living matter) and both a specific magnetite (a type of mineral) and sulphide particles that could have resulted from chemical reactions known to be important in Earth-based microbial systems. Once the data was released to the wider scientific community, the claims of life were mostly discredited as the evidence could have been produced by non-biological processes consistent with conditions on early Mars. This controversy has made scientists within the astrobiology community more wary of making claims that evidence for life has been found. As the Sagan Standard decrees: 'Extraordinary claims require extraordinary evidence.'[2]

Finding life beyond Earth might appear to be a frivolous endeavour undertaken by a privileged few. But think for a moment what it would mean if the life on Earth was the only life in the Universe, the only example of life in a possibly infinite

---

[2] Carl Sagan (1934–96) was an astronomer and science communicator who played a significant role in space exploration and astrobiology, as well as championing science accessibility to the public with series such as *Cosmos*. He'll come up a few more times!

Universe with infinite possibilities. It is, to me, the most important question we can ask ourselves as humans. Why are we here at all? Would the discovery of alien life change your views? Consider this as we go on this journey and see that although the search for life is a pursuit based in science it has far further-reaching philosophical implications. Science can often feel detached from emotion and humanity, but I hope you will see that the question of life is driven by both.

What Is Life?

Before we begin looking for life elsewhere, we need to establish what it is. Unfortunately, this is a question deserving of a whole library of its own and one that you will see does not have a particularly satisfying answer. A definition of life covers not only biological and physical properties but is also deeply rooted in philosophy. While philosophical principles are inevitable in this conversation – and in some fields of study researchers question whether we should even attempt to define what is living and what is not – we will ground ourselves in the science and our scientific understanding of life.

When we talk about life, we are referring to all living things – those on the macroscopic scale such as humans, animals, insects, trees and plants, and also those that inhabit the microscopic world, including bacteria and microorganisms. If I asked you whether your pet dog is alive, you would answer without hesitation, 'Well, of course they are!' And if I asked if your dining table is alive, you would give me a perplexed look that clearly says, 'No.' You know what is living and what is not. But how do you know?

The most widely accepted definition of life is of a self-organising, self-replicating, metabolically active system. Let's break down what that means and see what makes the living and non-living so different and, in some cases, not so different.

Self-organising

To enable the complex functioning of any living thing, organised systems are vital. Examples of these systems can be found in your dog. There are macro-systems, which are larger things we can see, like the respiratory system consisting of the lungs, diaphragm, oesophagus and mouth. Then there are micro-systems – very small things we need a microscope to look at – such as the endocrine system, which controls the production of hormones that give instructions to different parts of the body. For instance, when blood sugar levels rise after eating, hormones prompt the pancreas to release insulin, which lowers the levels to prevent damage. Within these systems, there are tissues composed of groups of specialised cells that have specific jobs. One type of specialised cell cannot do the job of a different type of specialised cell. Your dog started out as a single cell that divided into more cells. These cells then self-organised by moving to specific areas and differentiating into the appropriate type of specialised cell. Some formed muscle cells, others red blood cells. This process of specialisation and self-organisation happens in all living things and is what enables the construction of larger systems, like the respiratory system, where different groups of specialised cells work together to provide a function.

Self-replicating

Next, let's look at self-replication, or reproduction. When a dog reproduces, both parents' genetic information (their DNA) is passed on to the next generation. On a purely biological basis, all living things seek to continue their lineage through reproduction. Puppies are the result of sexual reproduction, which requires two parents. There are, however, other ways to produce offspring – through asexual reproduction, for instance. This is essentially one-parent reproduction, where an organism produces a clone of itself. The process has been observed in some plant species and even in animal species like starfish and Komodo dragons. To take a concrete example: if you leave a potato in your cupboard for too long, you will see it begin to sprout buds. These are new potato plants that are genetically identical to the parent potato. Asexual reproduction might seem to be far easier than sexual reproduction, but it does have flaws in that the lack of genetic variation slows down any adaptation within a species and makes the organism more susceptible to disease. I would hazard a guess that your table does not reproduce, but the tree it was made from may have passed on its genetic information before it was turned into a table. Presuming it is a wooden table that was, say, once part of an oak tree before being cut down, it may have reproduced in its previous life. Similarly, if your dog is old enough, it too may have reproduced in the past, even if it no longer can.[3]

---

[3] There are many examples of animals that cannot reproduce. For example, worker ants are all female, but only the queen ant can lay eggs.

Metabolically active

Metabolism is one of the most essential processes in all living organisms. It refers to the complex network of biochemical reactions that allows organisms to acquire energy, grow, reproduce and maintain their internal environment. Metabolic processes can be divided into two broad categories: **catabolism** and **anabolism**. Catabolism refers to the breakdown of complex molecules into simpler ones, which releases energy in the process. Anabolism, on the other hand, refers to the making (synthesis) of complex molecules from simpler ones, which requires the input of energy. This energy input comes from respiration, which is a series of biochemical reactions that occur within cells to convert energy (such as glucose) stored in food molecules into a form (like adenosine triphosphate (ATP)) that can be used by cells. There are two main types of respiration: aerobic respiration, which requires oxygen, and anaerobic respiration, which does not.

The latter usually takes place in smaller creatures in environments where oxygen may not be plentiful. Your dog, of course, aerobically respires just as your table once did when it was a tree. Without respiration, cells would not have the energy necessary to carry out metabolic processes such as muscle contraction, protein synthesis and DNA replication. These processes occur at a cellular level and lead us on to the discussion of the building blocks of life.

Carbon-based life

All life on Earth is carbon-based. No matter if the life in question is your dog, a tree or bacteria, all organisms are made up of the same fundamental building blocks. So, what goes into making a cell and what does carbon-based really mean?

There are six main elements that are necessary for life: hydrogen, carbon, nitrogen, oxygen, phosphorus and sulphur. These in turn form the four main biomolecules: nucleic acids, proteins, carbohydrates and lipids.

Nucleic acids are chains of pairs of nucleotides. The specific order and pairing of the nucleotides, which are themselves made up of a sugar, a nitrogen-containing base and a phosphate, can be used as a genetic fingerprint for every individual organism. The nucleic acids DNA (Deoxyribonucleic acid) and RNA (Ribonucleic acid) act as the instructions for life. These molecules contain the genetic information that controls the development and functioning of all living organisms. RNA is a single-stranded molecule, whereas the structure of DNA is the double helix you may be familiar with. DNA provides the stable storage of genetic information, while RNA plays diverse roles in gene expression and the transmission of genetic information. A human's **genome** (the entire set of DNA instructions) can be sequenced by defining the three billion base pairs that form it and determine that individual's unique characteristics, like eye or hair colour. These molecules come up again later, so it is important to understand the essential part they play in living things.

Proteins, meanwhile, perform a wide range of functions in living organisms, including accelerating chemical reactions, transporting molecules and providing structure to cells and tissues. They are made up of chains of amino acids that can fold into three-dimensional shapes to form functional proteins such as antibodies, which bind with any unwanted nasties like viruses in the body to eliminate them. The key elements of an amino acid are carbon, hydrogen, oxygen and nitrogen.

Next, carbohydrates, one of the main sources of energy for living organisms, provide fuel for metabolic processes. They are also important structural components of cells and tissues. Carbohydrates are composed of simple sugar molecules, such as glucose, fructose and galactose, all made up of carbon, hydrogen and oxygen.

Lastly, lipids are a diverse group of molecules including fats, oils and steroids, such as cholesterol, that make up cell membranes and serve as energy storage molecules. All six of the main elements for life on Earth may be present in different lipids.

Together, these four biomolecules provide the basic building blocks for the formation and function of living organisms. They are all carbon-based and it is the properties of carbon for which we should all be thankful. Carbon can form strong bonds with other elements and is incredibly stable; carbon-carbon bonds are some of the strongest in nature. It can create long chains of **polymers** – repeating patterns of molecules needed to form life essentials like DNA. Carbon is the fourth most abundant element in the Universe (by mass) and is widely distributed throughout the cosmos. Molecules with carbon-carbon or carbon-hydrogen bonds are known as **organic molecules**. All life on Earth is based on these molecule types, however, the existence of organic molecules alone does not necessarily mean there is life, a lesson the NASA scientists learned about ALH84001.

You might have noticed that 'carbon-based' wasn't included in the earlier, most widely accepted, definition of life. There are a few other related properties that are also commonly omitted and that deserve mention.

Response

Responding to the environment or a stimulus is important for any living thing. One such example of response for your dog is movement – your dog runs around and wags its tail. Movement is a necessary response to its environment and enables it to survive. For instance, when you feed it, your dog must move from one position to its food bowl and then chew and swallow its food. In a not-so-obvious type of movement, plants will turn themselves to face towards the Sun or another light source to increase the number of particles of light (**photons**) they capture for photosynthesis, the process by which they make their own food. These movements are all responses to external stimuli. Your table on the other hand? Unless you live in an animated film then I don't suspect your table moves of its own free will. So, is movement unique to life? Well, consider the natural phenomenon of a tornado. It moves across the surface of the Earth, but we would not say that a tornado is a living thing. In this case, the movement is simply a consequence of the prevailing winds pushing and pulling the funnel of destruction. It is essentially being moved, not moving itself, so its movement is not a response to stimuli. You can immediately see how this may be difficult to define, though.

Growth

Growth is an essential process for living things, but it is closely aligned to metabolism as it is a consequence of metabolic processes.

Your dog was once a puppy and, over time, grew to its adult size. It gained the energy to grow by eating food. On the other hand, your dining table might be extendable and 'grow' in size, but it does not grow on its own. Other non-living objects can display growth over time though. Under the right circumstances, including the addition of water, a crystalline structure will grow, thereby increasing the crystal or mineral's size. This is called autonomous growth. Although the structures are growing, they are not alive because they are simply acquiring more mass from external sources. In contrast, living organisms grow when they acquire energy through consuming food so their bodies can perform cellular and metabolic activities.

Homeostasis

Finally, living things have an optimum set of 'running conditions' and they self-regulate to keep within this fine set of parameters in a process called homeostasis. Just like a human, your dog must keep its internal temperature within a small range. When it becomes too hot, it will pant to cool itself down. Your table may be exposed to the burning heat of the Sun if left outside but it is not able to do anything about it, it must just accept its faded fate.

Evolution

Returning to things that are definitely living, NASA uses a definition for life based on Charles Darwin's Theory of Evolution, with its basis firmly rooted in the terrestrial life we know on Earth. It states: 'Life is a self-sustaining chemical system capable of Darwinian evolution.' Evolution, the process of change in and development of living organisms over time due to genetic drift and natural selection, deserves some consideration in any definition of life. It shows us how life has altered over billions of years and also how it has, in many ways, remained unchanged; the process of evolution is incremental – adaptations are seen over ages rather than generations and, as far as we know, the preconditions for life to exist on Earth have remained constant.

Genetic variation within a species arises from random mutations and 'recombination', which lead to differences in traits among individuals. A random mutation some 6,000 to 10,000 years ago is responsible for blue eyes in humans – until then, we all had brown eyes. Recombination is slightly different in that the genetic material exchanged by parents during reproduction is combined, resulting in the offspring having a mixture of traits that neither parent has. Natural selection, meanwhile, describes the mechanism by which individuals with beneficial traits are more likely to survive and reproduce, allowing those traits to be passed on to future generations. Over an extended period, an entire species evolves this new adaptation.

Humans evolved from the same common ancestor as chimpanzees, but they did not evolve from chimpanzees. In fact, humans, chimpanzees and all other life on Earth evolved independently from a microbe known as the last universal common ancestor (LUCA). LUCA was not the first life on Earth but is related to all current life on Earth. Due to its microscopic size and the challenging preservation conditions during its early existence, no fossil evidence of LUCA has ever been found, but its traits can be predicted using the genomes of its modern ancestors. The study of LUCA brings us closer to the origins of life and the latest research indicates that it was present on the planet just under 4.2 billion years ago. This is around the same time the earliest life on Earth is predicted to have existed – Earth itself is only 4.54 billion years old.

The tree of life (Figure 1) is a model and research tool used to explore evolution and describe the relationships between current and extinct life on Earth. Using

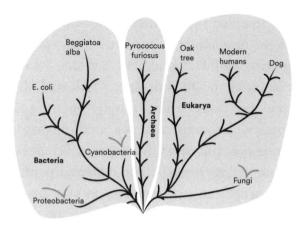

Figure 1. The tree of life is used to describe the relationship between life forms (current and extinct) on Earth. All life is categorised according to three groups, or domains: Bacteria, Archaea and Eukarya.

this framework, we can trace common ancestors, enabling us to see where we came from and, fundamentally, where all life came from.

There are three top-level groups, known as domains, in the tree of life: Bacteria, Archaea and Eukarya. Most of the life recognisable to us, and all the multicellular life we know of, is the latter – eukaryotic life. This includes humans, other mammals, plants, insects and the like, as well as some single-celled life, such as algae. Organisms of this domain are classified into a further seven major taxonomic ranks, with the next being kingdom, followed by phylum, class, order, family, genus and species. Modern humans are Eukarya, Animalia, Chordata, Mammalia, Primates, Hominidae, *Homo*, *Homo sapiens*. Up until the order rank (Mammalia in this case), humans share the same lineage as their pet dogs: Eukarya, Animalia, Chordata, Mammalia, Carnivora, Canidae, *Canis*, *Canis familiaris*. At each rank the tree of life subdivides into multiple further branches; the last shared common ancestor of modern humans and dogs was a shrew-like creature that existed around 100 million years ago.

There are thought to be millions of different species currently on Earth, with described species numbering only around 1.7 million.[4] As more species are

---

[4] Higher figures have been proposed but are usually downplayed due to the possibility of different descriptions being attributed to the same species.

discovered, other species are going extinct. Scientists predict that up to 150 species per day are lost, so it is nearly impossible for us to know how many species really exist on Earth at present. It is estimated that around 5 billion species have existed on our planet throughout its history.

When it comes to searching for extraterrestrial life, we would expect to see Darwinian evolution in any alien life. However, life elsewhere may not be as diverse as the life we see on Earth due to a lack of habitat variation or environmental pressures to adapt.

'Living fossils' are organisms that appear outwardly to have remain unchanged for millions of years. The term is, unfortunately, a bit of a misnomer. Although the creature may look the same as its ancient ancestor, on a cellular level it will have changed substantially through genetic drift. These living fossils (horseshoe crabs are one example) tend to be found in environments that have been stable for prolonged periods, where morphological (structural) adaptations would serve no benefit to the organism's survival. Put simply, if leaves on trees remain at a height easy for regular giraffes to reach and eat, giraffes with longer necks would be no more likely to survive to reproductive age and would, therefore, be no more likely to pass on their long-necked genes.

Viruses

Absent in the tree of life model are viruses. They are considered by some scientists to be living and by others to be on the edge of living organisms, blurring the lines between the two. Uniquely, they do not metabolise and require a living host to reproduce. They come in various shapes and sizes. One example is the enterobacteria phage T2,[5] which infects the bacteria *E. coli*. It is composed of a protein-coated

---

[5] Bacteriophage literally means 'bacteria eater'.

polyhedron head encasing its DNA and a tail encompassing a central sheath that leads to spindly tail fibres, giving the virus an insect-like appearance. This tail binds to the surface of the bacteria it is infecting and, once attached, the virus penetrates the cell and injects its DNA. The host then replicates the virus from this DNA and the new bacteriophages leave the cell. The process is known as **horizontal gene transfer** – the movement of DNA between organisms without the need for reproduction. Viruses don't themselves reproduce to pass on their DNA, they force it upon an already living being. They may be fundamental to the origins of life, though, as they can alter other viruses and living things, potentially spurring adaptations more quickly than through regular reproduction. It is possible that viruses have arisen independently several times on Earth and do not share a common origin like all other current life shares with LUCA.

We've explored the physical and chemical characteristics of what makes something living, but life may have a far wider definition. The problem with trying to form a definition of life is that we may be expecting a binary answer, but the point at which something goes from being non-living to living may not be so black and white; it may be part of a continuum. Some properties, including those already discussed, may be common among multiple different extraterrestrial life forms, while others may be unique.

Life is one of the most complex things in the Universe and, although we have identified many of its characteristics, it is unsurprising that there is not a definitive answer to the question 'What is life?' It is possible that we will never have a true definition. What we do know about life on Earth is a good starting point, though, and a way of furthering our understanding of potential life in the Universe is to delve into the origins of life and the evolution of our home planet.

Where Did Life Come From?

How was it that, after the planet first formed around 4.6 billion years ago, life developed on Earth? Is life an inevitable consequence of planetary formation? And is it inevitable that, given enough time, life will arise elsewhere? In our search for life in the Universe, it is important to consider these questions, the answers to which may shed light on how life might form in other environments in space.

The evolution of Earth

We have already covered the basics of the evolution of life on Earth but, when looking at how life emerged on our planet, we must also note the changes in Earth's environment over billions of years. Earth's history is divided into four eons, further split into eras, and each is characterised by significant geological and biological events (Figure 2). Rocks hold the key to Earth's early conditions, just as tree rings can reveal details of the changing atmosphere. Rock layers and their markings and structures record global events and the elements available to early life. The very earliest rocks have been eroded and recycled into Earth's core, so we don't have a complete picture of the planet's initial evolution.

The earliest eon, the Hadean Eon, spanned from 4.6 billion years ago to 4 billion years ago and is named after Hades (Greek god of the underworld) due to its hellish conditions, with surface temperatures reaching nearly 8,000°C

(14,400°F). Earth had just formed and the extreme heat was caused by the force of gravity compressing the planet, as well as intense bombardment by asteroids and volcanic eruptions, which resulted in a largely molten surface and magma ocean. It wasn't until around 4 billion years ago that the first solid crust formed, as Earth's surface cooled and solidified. While some research, including a 2001 study of minerals in Western Australia, has suggested that there may have been liquid water on Earth as far back as 4.4 billion years ago, it is difficult to confirm these findings without further work and more conclusive evidence to back them up.

Following the Hadean Eon came the Archean Eon, lasting from around 4 billion years ago to 2.5 billion years ago, and for this period we have a far more robust rock record to ascertain Earth's conditions. At this time, young Earth was shrouded in haze due to methane droplets in the atmosphere and there was little free oxygen (less than 0.001% of the oxygen in today's atmosphere).

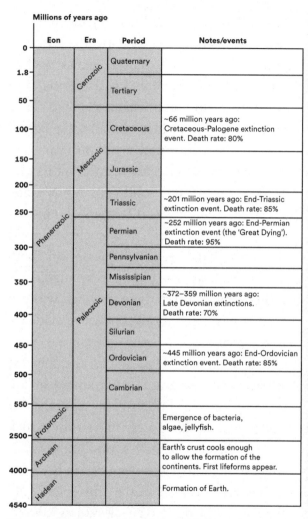

Millions of years ago

	Eon	Era	Period	Notes/events
0			Quaternary	
1.8		Cenozoic		
50			Tertiary	
100			Cretaceous	~66 million years ago: Cretaceous-Palogene extinction event. Death rate: 80%
150		Mesozoic	Jurassic	
200				
250	Phanerozoic		Triassic	~201 million years ago: End-Triassic extinction event. Death rate: 85%
			Permian	~252 million years ago: End-Permian extinction event (the 'Great Dying'). Death rate: 95%
300			Pennsylvanian	
350			Mississipian	
		Paleozoic	Devonian	~372–359 million years ago: Late Devonian extinctions. Death rate: 70%
400			Silurian	
450			Ordovician	~445 million years ago: End-Ordovician extinction event. Death rate: 85%
500			Cambrian	
550				
2500	Proterozoic			Emergence of bacteria, algae, jellyfish.
4000	Archean			Earth's crust cools enough to allow the formation of the continents. First lifeforms appear.
4540	Hadean			Formation of Earth.

Figure 2. The history of Earth is divided into four eons, which are subdivided by eras and periods.

The temperature of the planet is thought to have been similar to that of the present day, despite the Sun being fainter – a problem known as the faint young Sun paradox.[6] The warmer temperatures were likely caused by high concentrations of greenhouse gases, such as carbon dioxide, but this climate provided the conditions for the most vital liquid in the Universe to exist: water.

The importance of water in the search for life cannot be overstated. It is thought to be essential for any life and scientists often 'follow the water' when looking for environments in space that life might inhabit. Many chemical processes need a solvent, which is something that molecules can dissolve in. The molecules can then reassemble to form a different structure or be transported around an organism as part of a solution. For example, plants absorb nutrients that have dissolved in water through their roots in the soil. Water

---

[6] Stars evolve over time and the Sun would have been between 70% and 80% as luminous as it is today.

is hailed as the 'universal solvent', as it can dissolve more types of molecule than any other liquid. A water molecule has a positive charge at one end and a negative charge at the other end (like a magnet), making it a 'polar molecule' and allowing it to interact with other polar molecules, and other water molecules. It is the polar properties of water that cause raindrops running down a window to merge. Water molecules are relatively small, which allows them to move easily between other molecules, helping to dissolve substances.

Water can remain in its liquid state over a large temperature range (0–100°C or 32–212°F), meaning it can exist as a liquid in many more environments than other potential solvents. In its solid form, ice, it is less dense than liquid water, which is very unusual. As such, ice floats on top of liquid water, which is important for life in lakes and oceans. If ice sank, all the organisms at the bottom of a body of water would be frozen and during extremely cold periods in Earth's history all ocean life could have been wiped out. Water also has a high heat capacity, which essentially means it can absorb a lot of heat before its temperature starts to increase. In large bodies of water, this characteristic protects marine life from sudden temperature changes. Earth's proximity to the Sun is also paramount in ensuring liquid water could and continues to exist on the surface; not so close that it vaporises and not so cold that it freezes. The area around a star where water can exist in a liquid state is known as the **habitable zone** or the Goldilocks zone. This magical liquid will come up again and again in our search for life, so get excited about it!

But, back to the Archean Eon. Earth was covered in water for the majority of this time and, as such, was considered a water world.[7] Where there is water, there is life, and this eon is marked by the appearance

---

[7] A planet dominated by oceans and with few or no landmasses.

of the first life forms on Earth.[8] These were likely simple **prokaryotes**, which evolved in response to the changing environmental conditions on Earth's surface. The earliest known life and direct fossil evidence dating to 3.5 billion years ago was found in a rock sample in Western Australia. The team identified 'light-eating' or phototrophic bacteria, methane-producing archaea and methane-consuming bacteria. This diversity suggests that the first life must have developed much earlier to account for the evolution of such different life forms.

Next up: the Proterozoic Eon, which covers the period from 2.5 billion years ago to 541 million years ago. This eon is characterised by a series of important biological changes, including the advent of eukaryotic cells with more complex cellular structures and the evolution of multicellular organisms. The emergence of organisms capable of photosynthesis,

---

[8] This is when LUCA is suspected to have made an appearance.

like cyanobacteria (blue-green algae), had a profound impact on the evolution of life on Earth. The rise in atmospheric oxygen resulting from photosynthesis allowed for the evolution of more complex life forms with larger and more efficient metabolic pathways – more metabolism means more energy and therefore more growth. Photosynthetic organisms also had implications for the planet's climate, as the reaction of oxygen with methane in the atmosphere contributed to cooling, as well as a reduction in carbon dioxide levels.

Closer to surface level, the shape of landmasses on Earth has changed over time. Driven by plate tectonics, so-called supercontinent cycles – where continents come together and later break apart – occur over large timescales. The Proterozoic Eon is thought to have featured the first of these.

The Phanerozoic Eon (our present eon) is split into three eras, beginning with the Palaeozoic Era (541 million years ago to 252 million years ago). This era includes a pivotal moment in the evolution of life known as the Cambrian explosion, during which a remarkable diversification of life on Earth occurred that coincided with the appearance of the first fish, amphibians, reptiles and insects. Life began in the oceans in the Archean Eon but, by the end of this era, it had moved onto land and there were vast forests of plant life. The emergence of these more complex life forms was made possible by several environmental changes, including the development of land-based ecosystems and the evolution of more sophisticated respiratory and circulatory systems.

The end of this era saw what is commonly referred to as the 'Great Dying'.[9] It is considered to be the most severe extinction event in Earth's history, as it resulted in the loss of up to 96% of all marine species and 70% of land-dwelling vertebrates. Its cause is still debated, but it is thought

---

[9] Its formal names include the End-Permian extinction event and the Permian-Triassic extinction event.

to have been triggered by a combination of factors, including massive volcanic eruptions in what is now Siberia, which released enormous amounts of greenhouse gases and triggered global warming and **acidification** of the oceans. These sudden environmental changes led to a severe loss of marine habitats and a decline in oxygen levels, which in turn caused the collapse of marine ecosystems.

This mass extinction ushered in the Mesozoic Era, the age of the dinosaurs, approximately 252 million years ago to 66 million years ago. The period also saw the break-up of the supercontinent Pangaea, which greatly affected the distribution of life on Earth and paved the way for more varied habitats. It separated into two new landmasses: Laurasia (encompassing what we now recognise as North America, Europe and Asia) and Gondwana (South America, Africa, Antarctica, India and Australia). In these newly formed regions, a variety of unique plants (including the first flowering plants) and animals evolved independently, adapting to the specific conditions and resources available in each area. The isolation of these regions prevented species from freely migrating or interbreeding with species in other parts of the world, which resulted in the accumulation of genetic differences over time. Eventually, in a sign of the increased biodiversity on Earth, these differences became significant enough for organisms to be classified as separate species. As the continents moved further apart, vast oceans and seas formed between them. These new bodies of water, such as the Atlantic Ocean, the Indian Ocean and the Southern Ocean, created expansive marine habitats in which life diversified and adapted to varied conditions and ecosystems.

Furthermore, the break-up of the supercontinent significantly influenced climate patterns. The separation of landmasses altered weather patterns, ocean currents and atmospheric circulation, resulting in different regions experiencing variations in temperature, precipitation and other climatic factors. Some areas became more arid, while others became wetter or experienced distinct seasonal changes, and plant and animal species adapted and evolved to survive in their respective climates.

Finally, our current era, the Cenozoic Era, began after the Cretaceous–Tertiary (K–T) event – the extinction of the dinosaurs 66 million years ago. Following the catastrophic Chicxulub asteroid impact that erased the 'terrible lizards',[10] 75% of all plant and animal life was wiped out. The sudden absence of their former predators paved the way for the diversification and dominance of mammals, including the rise of primates and the evolution of hominids, and led ultimately to the emergence of modern humans.

Earth has experienced at least five big extinction events in its lifetime, during

---

[10] The word 'Dinosauria' is derived from Greek and is often said to mean 'terrible lizard'.

which almost all life was wiped out. What may seem like a disaster might actually be necessary in the evolution of life. Extinctions allow new life forms to develop and give less dominant species the opportunity to thrive, by removing predators or animals that hoard resources.

So, back in the Archean Eon, where did the first life come from? Aristotle believed that 'higher' life forms like humans had always existed, whereas primitive life spontaneously arose from decaying organic material, like leaves falling into muddy puddles. It's safe to say that this is incorrect but the notion that life may form by itself is not so ridiculous.

Abiogenesis, also known as spontaneous generation, is the hypothetical process by which life arises from non-living matter. It suggests that the first living organisms on Earth emerged from a combination of simple carbon-containing organic molecules and energy sources, such as lightning or volcanic activity, over a prolonged period of time. This process is thought to have occurred billions of years ago, when conditions on Earth were vastly different from what they are today.

In 1953, the Miller-Urey experiment attempted to recreate early conditions on Earth. Stanley Miller and Harold Urey created a closed glass apparatus, containing a mixture of water vapour, methane, ammonia and hydrogen gases – mimicking the proportions of these gases thought to be present around 4 billion years ago in a prebiotic (before life) Earth. A flask containing the mixture was connected to a series of tubes and a condenser, which allowed the gases to circulate and prevented the escape of any of the products. The mixture was heated and an electric spark was passed through it to simulate lightning. Analysis of the now condensed water a few days later showed that organic molecules were present, including the amino acids glycine, alanine and aspartic acid (remember amino acids are the building blocks of proteins needed for life). The significance of the Miller-Urey experiment lies in the fact that, although it did not produce life, it provided evidence that the basic building blocks of life could have arisen spontaneously from non-living materials under early Earth's conditions. More recent repeat experiments under conditions that – as a result of our improved knowledge of Earth's history – are more reflective of that time have produced similar results. These organic molecules were not living though, so how do we bridge the gap from chemistry to biology?[11]

Hydrothermal vents

The earliest life is thought to have been marine life and, as liquid water has been present on Earth for billions of years, it makes sense to investigate ocean habitats for its origins. As of 2023, though, less than

---

[11] If I knew the answer to this, I would be holding my Nobel Prize right now!

25% of the seafloor has been mapped. It is a difficult environment for humans to investigate, primarily due to the fact that we can't breathe unaided underwater – it's a bit of a limitation – and the extreme pressures. As you descend towards the depths of the ocean, the pressure increases. The pressure can reach over 1,000 atmospheres at the deepest points of the ocean, which would feel like 100 elephants weighing down on you. The vast majority of the seabed lies in the 'abyssal zone', between 3,000 m and 6,000 m (approximately 10,000 to 20,000 ft) below sea level. The depth of some oceanic trenches, however, has been measured at 4,000 m (13,000 ft) more than that. The Mariana Trench in the western Pacific Ocean, for example, is approximately 10,000 m (32,800 ft) below sea level at its deepest. Light cannot penetrate depths below 1,000 m (3,280 ft), so these environments are enveloped in darkness and you would expect them to be extremely cold and thus utterly inhospitable to life – but this is not the case. In the 1960s, new research into plate tectonics centred on the idea that heat from Earth's subsurface needs to be released in the same way that steam builds up and is released through the spout of your kettle as the water inside it boils. Hot springs on land, like the famous geysers in Yellowstone National Park in the US, work in a similar way. Below Earth's surface, groundwater is heated by magma, a geothermal process that causes the water to rise and, in some cases as in Yellowstone, to burst out in hugely energetic spouts of water and steam. If this was happening on land, then might it also be occurring on the seafloor, providing heat to an otherwise almost freezing region?

In 1977, scientists Tjeerd van Andel and Jack Corliss investigating the Galapagos Rift, a volcanic ridge near the Galapagos Islands in the eastern Pacific Ocean located some 2,500 m (8,200 ft) below sea level, made a startling discovery on board a deep-sea submersible called *Alvin*. They measured a temperature peak of

8°C (46°F) and in doing so pinpointed the location of the first known hydrothermal vent (image 2). This was the discovery the geologists had been hoping for, but they were not prepared for what else they encountered. Surrounding the vent were clusters of white clams that appeared to be thriving. Finding life at these depths had been so inconceivable that there was not even a biologist among the full crew of *Alvin*'s companion research vessel – no one thought there would be any need for one! Further excursions revealed the ocean floor was teeming with life; an otherworldly landscape home to communities of strange and spectacular life forms was uncovered. They came across fields of tubeworms (long white stalks with red tops), crabs and even a purple octopus feeding on clams. Soon after, biologists visited the site and detailed organism after organism that had never been seen before. This abundance of animals depended on the warm fluids flowing out of the seafloor, getting their oxygen for respiration from seawater.

But what were they eating? Bacteria (such as *Beggiatoa alba*) from around the vent from around the vent were found to be using hydrogen sulphide contained in fluids being expelled by the vents to generate carbon from the carbon dioxide dissolved in the seawater. This mirrored the process of photosynthesis in plants, except, instead of using sunlight, the microbes created their food source using chemicals – a process called **chemosynthesis**.

The discovery of hydrothermal vents reshaped our view of life and the conditions it can withstand, sparking interest among biologists searching for the origins of life due to the notion that early life relied on chemosynthesis, which had only previously been hypothesised, too.

Consequence of molecular cloud collapse

Early Earth conditions suggest that, with the right starting materials (organic molecules and water), life could evolve. But where do these ingredients come from? To answer this, we need to go back to the very beginning of time itself.

The Universe has been around for about 13.8 billion years, birthed by the Big Bang, where everything that has ever existed was just a tiny, extremely dense singular point before it burst outwards and expanded.

Around 370,000 years later, temperatures cooled enough for the first hydrogen and helium atoms to form – the lightest elements. And around 100 million years later, as the Universe continued to cool, the first generation of stars was forged from molecular cloud nebulae, giant clouds of gas and dust. Entire life cycles of stars passed until 4.6 billion years ago, where our local story begins. A molecular cloud became too massive and gravity caused it to collapse in on itself, beginning the process of star formation – just as in the case of the very first stars. Extreme gravity forced that material to come together to form our star, the Sun. The remaining debris from the cloud formed a protoplanetary disc, a dense, rotating disc of gas and dust around the Sun, some of which clumped together to form each of the planets in our Solar System (along with most of the other objects, including dwarf planets, some moons, asteroids and comets).

Models of molecular cloud collapse, supported by direct detections from observations of protoplanetary discs orbiting young stars elsewhere in our Galaxy, suggest that key organic molecules, including water, are generated in this process. We can assume, then, that all the ingredients for life were on Earth from its formation, or in the other astronomical bodies also formed from the protoplanetary material. This is where the habitable zone around a star is all-important too. The proximity of these bodies to the Sun will have been crucial in determining whether they could sustain liquid water.

Where did Earth's water come from?

One key question has been how Earth came to have its water, as it is thought that early Earth may have been too hot for it to form in such large quantities. There are many ideas surrounding the origin of Earth's water and it is likely that the reality lies in a combination. The first suggestion is that hydrogen and oxygen in young Earth's atmosphere reacted with the solar wind, producing some water. This argument, however, is not well supported. Another possibility is that Earth always had its water but that it was locked away in rocks until volcanic activity brought it to the surface and released it through volcanic outgassing, a process not unlike that seen in hydrothermal vents. This outgassing still takes place in the present day but at a much slower rate. One well-supported idea is that, during its formation, the planet may have gathered up tiny, hydrated grains of dust that existed in the protoplanetary disc. These would have melted on Earth once the planet heated up, releasing their water. Lastly, it is possible that the Solar System's asteroids and comets may also be responsible for delivering water to Earth.

Asteroids are mostly composed of rock and metals like iron or nickel but have also been found to contain water and are of particular note because this water matches that on Earth. You see, not all water is the same. There is 'light' water and 'heavy' water, with the latter literally being heavier than the former. The ratio of heavy to light water, which can be determined by the D/H ratio,[12] can tell us if water now in different places originated from the same source. Water ice was discovered in 2009 on the surface of the asteroid Themis, located in the asteroid belt between Mars and Jupiter, and has since been discovered on many more similar asteroids. The water on these asteroids, collectively known as carbonaceous chondrites or C-type

---

[12] The deuterium/hydrogen ratio.

asteroids, has been found to be heavier than the water on Earth so they cannot be solely responsible for its presence. However, researchers analysing samples of the S-type Itokawa asteroid returned by the Japanese Aerospace Exploration Agency (JAXA) in 2010 found a way to create light water from the particles. Dust grains from this S-type asteroid (containing silicates, a form of mineral), were bombarded with radiation replicating the solar winds that hit early Earth. Afterwards, the researchers found that significant quantities of light water had been produced. A combination of light water from S-type asteroid interactions and heavy water from C-types could answer the question of how early Earth came to have its water.

Comets, dirty space snowballs of dust, rock and ices, characterised by their shining tails, are thought to have deposited their water on Earth directly. Unfortunately, the D/H ratios calculated by scientists so far indicate that comets have too much heavy water to be the only source of Earth's water, but they have delivered when it comes to organic molecules.

In 2014, the European Space Agency's (ESA) Rosetta mission rendezvoused with Comet 67P/Churyumov-Gerasimenko (image 3), a comet orbiting between Mars and Jupiter that is thought to have originated in the Kuiper Belt – a doughnut-shaped region of icy objects orbiting the Sun beyond Neptune. The mission found not only water ice and water vapour on the comet but also glycine (an amino acid) and phosphorus, two essential components for life. The **panspermia** hypothesis suggests life may be transferred between planets and through space via bacterial spores in dust or on asteroids and comets. With an average temperature of -270°C (-454°F), pressure of just above 0 atmospheres (almost a vacuum) and high radiation levels from the Sun as well as cosmic events, the harsh environment of space is not thought to be conducive to life; it is a freezing, blood-boiling, radiation-sickness-inducing place.

Interplanetary, or even interstellar, transport of life could mean that life was brought to Earth from somewhere else. If that were true, then perhaps the origins of life on Earth are to be found in space. The long distances and timescales involved (to mention just two flaws in the panspermia hypothesis) mean that this is very unlikely to be the case. But it is important to understand that life can survive in the harsh space environment and the possible consequences of this on our exploration of potentially habitable astronomical bodies. In fact, there are a number of organisms known to have survived in some of these extreme environmental conditions.

Fungi are a type of eukaryote with some unique properties. Despite their appearance, they are far more like animals than plants as they do not photosynthesise and cannot produce their own food – they are their own eukaryote kingdom. Fungal cells are made from chitin, a substance not found in plant cells but present in the shells of some insects and crustaceans. For nutrition, cells branch out as a mycelium, a network of fungal threads or roots, which pushes enzymes for breaking down food outside its cells. This means it can survive in many different environments where food sources are very far away or in areas where nutrients are so hard to reach that only microscopic cells can access them. Fungi reproduce both sexually and asexually, expelling spores that can be transported via air or water. Experiments have shown they are able to survive the vacuum of space and withstand the Sun's radiation. Two fungi found on board the International Space Station (ISS), *Aspergillus* and *Penicillium*, were shown to be able to tolerate 200 times the radiation that humans can.[13] This indicates that some fungi could survive a journey through space over long distances on board a spacecraft or even on the surface of an asteroid or a comet. The superpowers

---

[13] They weren't taken into space on purpose – astronauts on the ISS are continuously battling mould growths!

of fungi have inspired further research to work out how radiation-resistant fungi could be used to protect electronics and even people on long space journeys. Given that evolution is prompted by the need to adapt to certain conditions, scientists are trying to determine why fungi have developed these characteristics – after all, there's no need for them to be so resistant to radiation on Earth. We do know, however, that fungi are Earth-born organisms, as they have a traceable common ancestor from around 1.5 billion years ago.

The potential for accidentally sending organisms from Earth to another planet and contaminating that environment are extremely important considerations for space agencies. It might lead to a false detection of life or the life brought from Earth could wipe out any possible alien life. If life is detected and it turns out that all life in the Solar System originated from the same source, it may be impossible to distinguish between alien life and the life that accidentally hitched a ride to another destination in the Solar System.

Acinetobacter is a bacteria that causes infections in humans and can survive clean rooms – controlled environments for the construction of spacecraft that are designed to filter out contaminants. There is speculation that some cells may have survived the journey to Mars on board NASA's *Curiosity* rover in 2012. It's unlikely, but impossible to know for certain.

Despite the potential for life to survive in space, it is abiogenesis that is the leading idea among scientists investigating the origins of life. While we do not fully understand how organic chemistry becomes biological and there may be no definitive distinction between non-living and living, our knowledge of life on Earth provides guidelines for our search for other life in the wider Universe, including what this life might look like.

What Would Life Look Like?

Since the earliest science fiction stories, humans have fantasised about what alien life might look like. You are probably familiar with the common 'grey', a biped of around four feet in height, with an enlarged head adorned with enormous eyes – the classic alien type supposedly found at the infamous Roswell 'crash site'.[14] This view of

---

14 The debris of a military balloon that crashed in 1947 in Roswell, New Mexico, US, was speculated to be the remnants of an alien flying saucer decades after it had been recovered.

aliens may seem wholly unimaginative, but, despite their unhealthy complexion, greys are not too far removed from humans. The idea that life elsewhere would resemble the life we find here on Earth is not implausible. In fact, it could very much look like something we're familiar with.

As we've seen, life evolves according to its environment. An adaptation suited to one environment may also be suitable for another, even if the environments are drastically different. A fitting example of this on Earth is in the development of the eye in both humans and octopuses. Essentially, both species have the same type of eye, known as camera eyes. The pupils allow light into the eye, changing size depending on whether more or less light needs to be let in, just like the aperture in a camera. While both octopus and human eyes utilise the same physiology, they evolved at different times. Looking back through the tree of life, humans and octopuses last shared a common ancestor – a flatworm that trawled the ocean floor – around 500 million years ago. That ancestor did not have the eyes humans and octopuses share today, meaning the camera eye evolved independently in both species, despite one living on land and the other underwater. In a similar way, if alien life evolved with its environment, it's possible that it may look remarkably like life on Earth, or at least have similar features. It is therefore reasonable to suggest that life which originated on a water world may have gills and fins like fish, whereas life that emerged on a desert planet may be characterised by features common to lizards or scorpions. Over time, as environments change on other planets, any life would be forced to evolve to survive, so it may vary hugely across the planet just as life does on Earth.

Marine life is extremely diverse and there is much that we have yet to discover. If you had never seen a giant squid and someone told you it was an alien found in the seas of a distant planet, you may well believe them because it is such a strange-looking creature. Many animals have adaptations we might describe as 'alien', because human and mammal adaptations are our 'normal', our baseline. Almost all cephalopods, the class containing squid and octopuses, can change the colour of their skin by a process known as metachrosis. This is a camouflage technique used when the animal feels threatened or to attract a mate. Cephalopods can do this at will and can even alter the texture of their skin. If we had never discovered animals on Earth that could behave in this way, we would be astounded to find alien life doing so.

Similarly, life elsewhere may perceive its surroundings entirely differently to humans. Animals like bats use echolocation to 'see' in the dark. The flying mammal emits sonar (a high-frequency sound), which bounces off objects around it. The soundwaves return to the bat, enabling it to build a picture of the size and shape of things in its immediate environment. Other organisms can see in different wavelengths of light; bees and butterflies can see in ultraviolet light (UV) and some species of snake can see in infrared. These wavelengths are outside the visible range for humans and it is difficult for us to understand the impact this different view of a world may have on behaviour and appearance for both terrestrial and alien life.

Of the countless species currently in existence on Earth, it is unsurprising that we find some life similar to us and other life that differs extensively. Of course, the life we are most familiar with is human life, but despite our intelligence and advancements we are incredibly fragile creatures. Humans need a specific set of conditions to survive in – to be entirely honest, we're terrible at surviving. Although we inhabit every continent,[15] humans can really only live in temperatures between 4°C and 35°C

---

[15] While there is no permanent population in Antarctica, scientists spend extended periods there conducting research.

(39 to 95°F), pressures of approximately 50% to 250% the pressure at sea level (1 atmosphere)[16] and need a consistent supply of oxygen among many other things. While technology has allowed us to live in a wider range of conditions, there are endless examples of other life that isn't quite so particular.

Extremophiles

From the peaks of mountains to the depths of oceans and the most arid of deserts, life can be found in every area of the world. What is most interesting about the life we find in these seemingly inhospitable environments is that it isn't just surviving, often it is thriving. There's a group of miraculous organisms known as **extremophiles** that really do thrive under extreme conditions.

---

[16] This doesn't include divers, as they use breathing apparatus with which they can withstand pressures up to 100 atmospheres!

One such extremophile is *Pyrococcus furiosus*, the name of which means 'rushing fireball'. It is a type of archaea, discovered in the 1980s on Italy's Volcano Island. You may be noticing a bit of a theme here – this little thing likes it hot, really hot. The sands on the beaches this organism inhabits regularly reach temperatures in excess of 100°C (212°F). It actually prefers these high temperatures and flourishes – this is what makes it an extremophile, a hyperthermophile specifically, meaning 'lover of very hot things'. Other types of extremophile include acidophiles (which love acidic environments) and halophiles (high salt concentrations). As we become more adept at exploring extreme environments, we will likely discover even more.

Tardigrades are extremophiles that you may have heard of before, although, strictly speaking, they're not actually extremophiles. These eight-legged, near-microscopic invertebrates have been popularised in recent years due to their endearing nicknames ('water bears' or 'moss piglets') and squishy appearance (image 4). They are not adapted to thrive in any particular extreme environment, much like our radiation-resistant fungi. They are, though, of great interest to scientists as they are almost impossible to kill and can survive in a range of extreme environments. So, what can you do to tardigrades without sending them to their doom? Well, you can freeze them, boil them, crush them with extreme pressures, expose them to the vacuum of space and heavily irradiate them – pretty much anything goes. In fact, tardigrades can be dried out (or desiccated)[17] into a state of suspended animation, where their metabolic rate – the amount of energy an organism uses to sustain basic functions – drops to just 0.01%. Using such little energy enables the minute creatures to survive decades in this

---

[17] Yes, like the coconut.

state and, once rehydrated, they spring back to life![18]

In 2019, the Israeli private company SpaceIL sent the *Beresheet* spacecraft to the Moon with a number of tardigrades on board for experimental purposes. Unfortunately, the unmanned spacecraft suffered a fatal crash on, or rather into, the Moon's surface, leaving some concerned that the tardigrades may contaminate the lunar environment. Ultimately, researchers concluded that the impact speed of around 500 km/h (310 mph) was simply too high for them to have survived, so perhaps tardigrades aren't quite as indestructible as they seem!

We've found other resilient organisms too. Scientists from Ghent University investigating extremophiles heated *Bacillus amyloliquefaciens* bacteria to 420°C (788°F). Despite evidence of DNA damage occurring during the experiment,

---

[18] Their average lifespan is usually between three months and two years.

once cooled, the bacteria successfully reproduced, suggesting they have a very effective DNA repair mechanism. As with tardigrades, this bacterium is not an extremophile, but exemplifies the hardiness of individual species and the spectrum of parameters under which life has been found to survive.

Alternatives to carbon-based life

Despite the immense variety of life on Earth, as we've already established, every living thing we know of shares a few basic properties: a building material, carbon; a solvent for reactions to take place, water; and a set of instructions, RNA or DNA. Earlier we discussed the brilliance of carbon and its unique ability to form long and complex molecules, but what if life was based on a different element altogether? Other hypothetical types of biochemistry, such as silicon-based life, have been proposed in place of the carbon-based life on Earth. Silicon has comparable properties to carbon; they are close together on the periodic table. While silicon is not as versatile as carbon and can't bond with quite such a diverse range of other elements, it is still capable of forming the long chains that we associate with life. It's also not as abundant as carbon in the Universe, as it is a heavier element and heavier elements require more energy to form.

Similarly, non-water-based solvents such as ammonia and methane have also been proposed as building blocks for life elsewhere in the Universe. Both substances are liquid at far lower temperatures than water – ammonia, for example, is a liquid between -77°C and -33°C (-107 to -27°F). Given that the freezing point of water is 0°C (32°F), the potential for other solvents to remain liquid under a wider range of conditions greatly increases the number of potential habitable environments. It is worth noting that there is no evidence for life based on any of these alternative biochemistries so far, but as we cannot be sure that Earth-like life is the only type of life in the Universe it is important to keep an open mind. If life was based on different biochemistry, there is no telling what it would look like or how it might behave!

Science relies on observation and experimentation. Usually, experiments undergo rigorous testing with many samples and repetitions to increase reliability. When it comes to the topic of astrobiology, we only have one example of life in the Universe: all life on Earth. Astrobiology is unique in that scientists also hypothesise and search for evidence of life using the knowledge we have developed on the basis of what we see around us.

In truth, other life may exist all around us – we simply can't detect it. This is because humans are three-dimensional beings in a four-dimensional Universe: three dimensions of space (up and down, left and right, backwards and forwards) and one dimension of time. According to complex quantum mechanical theory, the Universe exists in as many as 10 or 11 dimensions. Don't even try to imagine it, you really can't. There is nothing to say that life couldn't exist in a different dimension to our own, we just wouldn't ever know about it. And, if it did exist, would it even matter? We would never be able to comprehend its existence, after all. Food for thought for the philosophers among us!

So, what exactly are we looking for? And what do we expect to find that will indicate life?

What Are We Looking For?

In our everyday lives, we encounter signs of life without directly observing the living organisms themselves. If you saw a paw print in some mud, you would know a dog (or something similar) had recently been there. If you happened upon branches stacked across a river to make a dam, you would suspect the presence of a beaver. In the field of astrobiology, these signs are referred to as **biosignatures** and they encompass various characteristics, elements, molecules, substances and features that serve as evidence for past or present life. Importantly, they must be indicators that cannot be produced without the presence of life. This can be a bit of a grey area; we have discussed the potential existence of life different to life on Earth, which may present a biosignature we are not familiar with as a result. Although we don't expect to find anything as obvious as paw prints or animal-made constructions on other planets, we may find fossils preserved in rocks, organic molecules synthesised by living organisms, or differences in atmospheric or water chemistry that could indicate life.

Morphological biosignatures are identifiable structures or formations left behind by living organisms. Microbial communities in shallow waters can create layered mineral structures called stromatolites. These structures have been found in some of the earliest rock records of life on Earth from the Archean Eon. If similar layered minerals were discovered on another planet, they could be considered potential morphological biosignatures.

Chemical biosignatures cover a wide range of ways in which life can leave its chemical mark on rocks, bodies of water and even atmospheres. They include the biomolecules we discussed when defining life (lipids, carbohydrates, nucleic acids and proteins) and the products of chemical processes in living things.

Then there are molecules known as chiral molecules. These are molecules that have a 'handedness'. In the same way that you have a left and a right hand that are mirror images of each other, chiral molecules are mirror images of the same molecule. Biological material tends to have an equal mix of 'left' and 'right' versions. In humans, glucose is a chiral molecule and we would expect to find equal numbers of 'left' and 'right' glucose molecules. In non-living substances, there is usually a preference for 'left' or 'right' chiral molecules. If they were found elsewhere in space, we could use the presence and proportions of these molecules to distinguish between living and non-living.

Scientists can also analyse the ratios of isotopes (different versions) of chemical elements. This is similar to how we look at the D/H ratio when searching for the origins of water on Earth. Life as we know it tends to favour the use of lighter isotopes, as they provide more energy for metabolic processes and growth. Measuring the ratio of lighter isotopes to heavier isotopes in samples from nature and finding a higher proportion of lighter isotopes could serve as a biosignature.

So how do we actually look for these biosignatures? Well, we can do a fair amount of searching with ground-based telescopes here on Earth. Each iteration of the telescope has furthered our knowledge substantially. It all began with the optical telescope back in the 1600s, invented either by Galileo Galilei (1564–1642) or a German-Dutch spectacle-maker named Hans Lippershey (c.1570–1619) – the jury's still out on that one! It was around this time that Galileo observed the first moons of another planet – the four largest moons of Jupiter, the Galilean moons. Not only did this expand our knowledge of the objects that lie within our Solar System, but it also reshaped physics: Earth could not be the centre of the Solar System if other planets had moons orbiting them too. Although true, this then revolutionary idea saw Galileo persecuted and imprisoned for the rest of his life. Over the centuries, optical telescopes grew in size, enabling the discovery of objects far further from Earth. In 1930, American astronomer Clyde Tombaugh (1906–97) discovered the dwarf planet Pluto using an astrograph – a telescope used for taking wide-field photographs of the night sky on photographic plates. Optical or visible light telescopes are still used today for astronomical research.

However, the Universe is teeming with electromagnetic (EM) radiation. Light exists on a spectrum and you are already familiar with many of the different types,

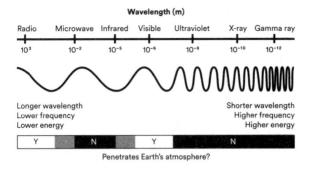

Figure 3. The electromagnetic (EM) spectrum. Only some wavelengths of EM radiation penetrate Earth's atmosphere.

or wavelengths, of radiation, of which visible light and infrared radiation form just a small portion (Figure 3). By using telescopes that can 'see' in different lights, we can build a much more complete picture of space and all its environments.

Spectroscopy

As we know, geologists have learnt much about Earth's formation and its evolution by studying its rock layers, which keep a record of events that occurred locally or globally and allow us to build up a picture of what was going on at any given time. Similar techniques are used in the search for evidence of life elsewhere.

When EM radiation reacts with a substance, a unique fingerprint showing the radiation the material absorbs and emits can be determined using a process known as **spectroscopy**. This can reveal the substance's composition. Think of it like a well-trained ear determining which instruments are playing during an orchestral performance. It can hear the entire composition and pick out whether the violins or violas in the string section, or both, are producing the sound. Just as the composition of a song can be identified, the composition of a rock or other substance can be picked apart too. Chemical signatures can be detected in atmospheres using the same spectroscopy techniques. Light from a planet's star acts as the laser, passing through a planet's atmosphere, which in turn absorbs and emits different wavelengths – its signature spectrum. Telescopes can identify gases in planetary atmospheres, such as methane or an abundance of oxygen (which, as two of the components in some of the key biomolecules for life on Earth, could indicate life), using this method.

While direct observations from ground-based and space telescopes help us search for biosignatures, explorative space missions can offer far more insight. Most first visits to other worlds are flybys, one-time passes, performed by spacecraft with instrumentation sufficient to study an astronomical body's general conditions and take images. Orbiters are spacecraft in orbit around a body and function in the same way as the satellites orbiting Earth, which send signals such as GPS for navigation or image the planet's surface to monitor environmental conditions and events.

Decisions to make sure the design and engineering of future craft are fit for purpose on can be made based upon the details from these initial missions. It's like sending out a scout to make sure you're prepared for the journey ahead. More advanced spacecraft, such as specially adapted landers and rovers, can then be sent to these environments to take measurements from the surface. They're usually equipped with a multitude of instruments that allows them to perform in-situ data collection, providing different and complementary information to that gathered by distant orbiters or other probes. From the surface of another planet, for example, robotic explorers can directly capture the gas of an atmosphere and determine its composition through mass spectrometry. This is similar to spectroscopy but uses the atomic mass of different elements to distinguish between them and, in turn, the composition of a substance. NASA's *Insight* (2018) lander is fitted with a seismometer, which listens out for Marsquakes (earthquakes but on Mars!). This device detects seismological activity (quake-related movement) on the planet, which can be indicative of geological activity, something we know to be important in a life-bearing body. Combinations of other instruments help scientists form a coherent understanding of a planet's environment: magnetometers measure the magnetic field, laser altimeters tell us about the topography and features on the surface, and particle detectors give insights about solar wind and cosmic rays. More obvious signs, however, could be directly observed.

Artificial signs

The ISS orbits Earth at a height of 408 km (254 miles). In astronomical terms, considering the Moon is on average 384,000 km (238,855 miles) from Earth, the ISS is incredibly close to our home planet. The astronauts on board can observe it 24 hours a day from the Cupola module, an observation deck with seven windows. Their relative proximity to Earth, however, limits their view of the planet to around 3% at any one time. Despite having a somewhat 'zoomed-in' view of the planet's surface, artificial constructions are rarely visible to the eye, except at night, when city lights can easily be seen. By contrast, astronauts involved in the Apollo missions (1961–72), who, in reaching the Moon, travelled further from Earth than any other humans have so far, were able to see the planet in its entirety from the window of their spacecraft. At a distance this great, no artificial structures on the planet's surface could be made out with the naked eye. Given the distance between Earth and all other astronomical bodies in the Solar System, it is extremely unlikely that we will be able to make out artificial structures (if there are any) on another planet or moon, even with the aid of an optical telescope, unless they are extremely large and stretch far across its surface.

Lessons from the history of the observation of Mars warn us of making assumptions about artificial features. Nineteenth-century astronomers wrongly thought they had observed canals channelling water from the ice caps at the poles towards the arid equator on the planet's surface (where it was thought civilisations were more likely to exist due to the region's more temperate conditions). Having first been described by the Italian astronomer Giovanni Schiaparelli (1835–1910) in 1877, the claims were firmly disproved by Edward Walter Maunder (1851–1928) of the Royal Observatory in Greenwich, who demonstrated that earlier observers had essentially 'connected the dots' of natural features across the Red Planet's surface like a giant dot-to-dot puzzle. His work invalidated any evidence for artificial surface features and thus the existence of Martians – it was just an optical illusion.

Habitability in the Solar System

Hanging in the darkness of space, our pale blue dot appears to be the perfect place for life to exist and thrive, so what makes Earth such a habitable utopia? Our planet has a unique setup within the Solar System. Most importantly, it is firmly in the habitable zone and able to sustain copious quantities of liquid water on its surface for long periods of time. This is due to a number of factors.

Firstly, orbiting the Sun at an average distance of 150 million km (93 million miles), Earth receives an optimal amount of heat and the stability of its orbit helps to regulate temperature fluctuations too. It has a rotational period of just under 24 hours,[19] which means there are only minor temperature differences between day and night across the planet. As a result, the temperature of the planet's surface is maintained between -25°C and 45°C (-13 to 113°F), averaging at 14°C (57°F), allowing liquid water to exist. Earth is tilted by 23.4 degrees on its axis and it is this tilt that gives the four distinct seasons. The half of Earth tilted towards the Sun experiences its summer while the other half experiences its winter – the northern hemisphere receives most sunlight between June and September and the opposite is true in the southern hemisphere. The seasons help

---

[19] 23 hours 55 minutes is the length of a sidereal day, which is how long it takes Earth to spin once on its axis. A regular day to us, known as a solar day, lasts 24 hours. This is the time it takes for the Sun to be at noon (its highest point in the sky that day) until noon the next day.

to give rise to unique climates across the planet, adding to the diversity of life.

Earth's atmosphere is composed of mostly nitrogen and oxygen and it is the right density to help stabilise the temperature differences caused by the planet's rotation and to create pressure on its surface. Pressure at sea level is 1 atmosphere. This is the pressure on land that most humans experience in their everyday lives, although there are permanent settlements at altitudes of up to 5,100 m (16,700 ft) above sea level, such as La Rinconada, Peru, where the air pressure is low enough to cause altitude sickness and other health issues due to lack of oxygen. Local people who have grown up under the conditions generally fare much better than tourists. In mountaineering, the 'death zone' is considered to be any region above 8,000 m (26,000 ft), where the pressure drops to around 35% that at sea level, the atmosphere thins and oxygen levels are insufficient for humans to survive unaided for any length of time.

Keeping the atmosphere safe is Earth's magnetic field. In combination, the two protect living things from too much solar radiation and cosmic rays from space. The atmosphere also shields us from small meteors, which burn up in it.

Due to the enormous amount of energy involved, a planet is ridiculously hot when it forms. Before the collapse of a molecular cloud to form a solar system, every particle of gas and dust inside it is moving in space. As the cloud condenses, the movements of all these particles get condensed too, which causes the cloud and, subsequently, the new planet to spin. The phenomenon is known as 'conserving angular momentum' and the same mechanism acts on an ice skater spinning with their arms out, who then brings them in tighter to rotate more quickly. So, as a consequence of the cloud collapse in our own Solar System, Earth and the other planets spin or rotate, giving them not only night and day, but also an electromagnetic field. The magnetic field is generated by what is known as the 'dynamo effect', whereby the conductive materials in Earth's core (iron and nickel), surrounded by molten rock and encased in solid rock layers, rotate around one another and produce electric charge. The circulation of this electrical current then produces a magnetic field. The magnetic field extends into space and forms a region called the magnetosphere, which acts as a protective shield, deflecting most of the charged particles from the Sun (the solar wind). This field is vital in protecting life on Earth from harmful radiation from space that would otherwise reach the surface and damage anything in its path. It also prevents the solar wind from stripping away Earth's atmosphere and therefore its water.

Although Earth's atmosphere is protected by its magnetic field, the planet's gravity is what allows it to hang on to it. A body's mass dictates the strength of its gravity, which is what keeps us all on the ground and prevents us from flying off into space every time we jump. The strength of gravity on Earth is 9.81 m/s² (32.17 ft/s², known as 1g), which allows for normal bodily functions in land-dwelling species like humans. Long-term skeletal deterioration, reduced red-blood-cell production (causing fatigue and dizziness) and many other physiological issues have been observed in astronauts who have experienced near zero-gravity in the microgravity environment (in which objects appear to be weightless) of the ISS.

Lastly, Earth has regenerative cycles. Geological activity, such as plate tectonics and volcanic eruptions, ensures a dynamic and changing environment. When living organisms die, their biological material gets recycled into the earth providing nutrients for new life.

Of the eight planets in the Solar System, the inner four are known as the terrestrial planets and the outer four are the giant planets (Figure 4). Terrestrial means Earth-like, but in this context we really only

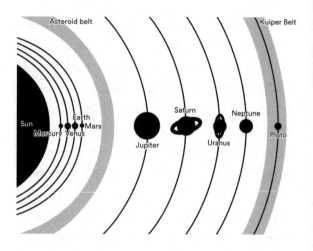

Figure 4. The Solar System, including the position of the asteroid and Kuiper belts.

mean solid and rocky when compared to the giant planets, which are made of gases and ices and have no solid surface. Looking for life on the planets most like Earth seems to be the best place to start, so let's begin at the closest planet to the Sun.

Mercury

Mercury orbits the Sun at an average distance of 58 million km (36 million miles).

That may seem far away, but it is no surprise that the side of Mercury experiencing daytime (and illuminated by the Sun as the planet rotates) is extremely hot, given that it is so close to a star with an outer atmosphere hotter than 1 million °C (1.8 million °F).[20] Unlike Earth, the planet lacks a thick atmosphere to help regulate the global temperature. Temperatures soar up to 427°C (800°F) on its sunlit (daytime) side and, without sufficient atmosphere to retain this heat, plummet to around -173°C (-279°F) on its dark (nighttime) side. Such extreme swings would necessitate adaptations in life to equip it both for intense cold and heat. The absence of a substantial atmosphere around Mercury offers minimal protection against the Sun's radiation and solar wind, as well as any meteorite impacts. The planet's surface is heavily cratered and scarred from numerous impacts by asteroids and comets. Any water present on the planet

---

[20] Stars have atmospheric layers just like planets.

would either quickly evaporate due to the scorching temperatures or remain frozen in permanently shadowed regions (PSRs) near its poles that never receive direct sunlight. This lack of liquid water is a significant obstacle for the existence of life.

Although we have collected data about the planet from orbiters and flybys, no missions have ever landed on its surface, so it is largely unexplored. The smallest planet in our Solar System is thought, by all accounts, to be a barren rock that has never been able to sustain liquid water on its surface. A scathing review of the prospects for current life, but what about past life? A 2020 study conducted by researchers at the Planetary Science Institute in the US suggested that certain regions in Mercury's subsurface may once have been suitable for prebiotic chemistry or even simple microscopic life forms. Its focus was Mercury's chaotic terrain, which was initially thought to have been formed by seismic disturbances from the impact responsible for the Caloris Basin, a huge crater around 1.5 times the length of the UK. Analysis indicated that the rugged terrain was created by the removal of large amounts of volatile elements (substances that evaporate easily) from Mercury's upper crust. The implication was that this formerly volatile-rich crust might once have been thicker and contained water or water ice, although there is no definitive evidence for this. The findings raise the prospect that some subsurface regions of Mercury may have once been hospitable to life. So, despite Mercury's current desolate conditions, we can't dismiss its past astrobiological potential. But, until we send missions to collect more data, we won't be able to draw any more detailed conclusions.

Venus

Venus is, ostensibly, a hellscape that would kill a human in more ways than I'm willing to spend the time describing, but let it be known that Earth's 'evil twin' deserves this nickname. At over 460°C (864°F), the planet's surface is hot enough to melt lead within seconds[21] and the surface pressure is the same as the pressure 1,000 m (3,280 ft) underwater on Earth – the equivalent of 100 cars stacked on top of you. Understandably, it is home to no liquid water, despite being at the inner edge of the Sun's habitable zone.

Due to Venus's rather uninviting surface conditions, it has been difficult to send missions there. We know much less about the planet than we do about our other planetary neighbour, Mars, but there have still been findings that have piqued our interest. Early missions to Venus between the 1960s and 80s, such as the Soviet Venera programme and NASA's Mariner and Pioneer Venus missions, revealed a planet characterised by extreme temperatures, a dense carbon-dioxide atmosphere and

---

[21] On Earth the melting point of lead is 327.5°C (621.5°F), but melting and boiling points vary with pressure.

corrosive sulphuric-acid clouds. Its thick atmosphere created a so-called runaway greenhouse effect, resulting in the high surface temperatures. However, after Venus's initial formation, when the Sun was fainter, the planet may have been able to sustain water for a period of time. Trace amounts of water vapour detectable in its atmosphere today indicate that water was present at some point, but for how long is very much up for debate among scientists. Some estimate that the planet was covered in liquid water for up to 2 billion years, while simulations by scientists at the University of Chicago in 2023 produced a figure of 900 million years. As water is heated, it breaks down into hydrogen and oxygen (its component parts), so you would expect there to be evidence of this oxygen remaining in the atmosphere or perhaps absorbed into rocks on the planet's surface. The scientists could not explain the lack of oxygen on present-day Venus with their simulations and used this as evidence that the planet may have had far less water on its surface than previously thought. There is a lack of water on the surface, but what is going on above it?

In 1995, the Hubble Space Telescope imaged Venus in UV. Scientists noticed dark patterns in the clouds, which signified (alongside data gathered from earlier flybys and probes) that something was absorbing this light – maybe something biological. In 2021, data from ground-based telescopes (JCMT and ALMA[22]) suggested high levels of phosphine in Venus's atmosphere. Researchers theorised that life could be responsible – geological processes on Venus could not fully explain the measurements taken. On Earth, anaerobic bacteria produce large quantities of phosphine (as do some industrial processes). While this link may not be sufficient evidence to confirm the presence of life in the clouds of

---

[22] The James Clerk Maxwell Telescope in Hawai'i and the Atacama Large Millimeter Array in Chile.

Venus, it is an interesting proposition when it comes to the extreme environments life could inhabit.

Although the temperature of Venus's surface is extremely high, the temperature around 50 to 60 km (31 to 37 miles) above it in the cloud layers drops to between -20°C and 60°C (-4 to 140°F), a much more reasonable temperature for life. The pressure here is far lower, too, around the same as at sea level on Earth. It's a region that appears more conducive to biological processes. Despite this more temperate environment, the Venusian clouds have an extremely high concentration of sulphuric acid – they'd be toxic to inhale and incredibly corrosive to living organisms as we know them. There are acid-loving extremophiles (acidophiles) that live around deep-sea vents on Earth, so it is not beyond the realms of possibility that an organism adapted to a sulphur-heavy, highly acidic environment could exist. However, the concentration of sulphur in the clouds of Venus is far higher than that in the regions where these organisms are found on Earth, so conditions may well be too extreme for any life.

Mars

Next up is our other planetary neighbour, Mars, who you might think gets a little too much attention. I assure you, though, the buzz around the Red Planet is not for nothing – there's good reason it's of such great interest to scientists looking for life. Second to Earth, Mars is the best explored planet with over 30 successful missions (including flybys, orbiters, landers and rovers) to the planet. Many of these missions have found evidence of liquid water both past and present, from apparent flows originating at the polar caps to huge reservoirs lurking beneath the surface.

Mars is, on average, 228 million km (142 million miles) from the Sun, which is only around 78 million km (48.5 million miles) further than Earth. I say 'only', as in astronomical terms that's barely one row further back in the theatre and places Mars at the very outer edge of the habitable zone. This means the amount of sunlight and heat received by Mars is enough to give the planet an average global temperature of -62°C (-80°F). Pretty chilly, but in the summer the temperature around the equator can reach highs of 20°C (68°F). At times, therefore, the temperature on the planet's surface could enable it to sustain liquid water – in certain regions and for a very short while. Now, the recurring theme so far throughout this book is liquid water; all life needs it and it excites astrobiologists like nothing else.[23] But it is not only temperature that dictates the liquidity of water – pressure and radiation also play an important role. If the pressure is too low or radiation levels are too high, liquid water cannot exist – it will turn to vapour.

---

[23] Perhaps second only to an anagram.

Today, Mars appears to be an arid, uninhabitable dust bowl. Covered in iron oxide (rust) and with an atmosphere 1% as dense as that of Earth, it is not surprising that we don't see life roaming the planet's surface. The Red Planet hasn't always been this way, though: it was once covered in liquid water. It had oceans, lakes and rushing rivers. It would even have had a water cycle, with clouds of rain emptying themselves over the landscape, topping up bodies of water that lost their volume to evaporation as the water was heated, before condensing back to liquid and cooling to forming clouds.

The presence of sedimentary rock formations, discovered by NASA's *Curiosity* (2012) rover, is consistent with ancient dried-up streams and lakes. These rocks contain rounded mineral grains, suggesting they were transported by water – like pebbles on an Earthly beach whose hard edges have been removed over the years by the sea. *Curiosity* also confirmed the existence of valleys and canyons on Mars similar to those formed by rivers on Earth. One of the most prominent examples is Valles Marineris, a system of canyons around the width of North America and over 4,000 km (2,500 miles) long, up to 7 km (4 miles) deep and likely created through erosion by water.

But let's get back to all that water: what happened to it? Well, the short answer is that Mars lost its atmosphere and, with it, its water – a consequence of the Red Planet's size (half that of Earth), which caused it to cool more quickly than Earth (further spurred on by its added distance from the Sun). Its geological history bears similarity to Earth and the planet could once have been a very habitable place.

Mars formed around 4.5 billion years ago, at the same time as Earth and other Solar System objects. Initially, it experienced intense volcanic activity (much like our planet), creating the Tharsis volcanic plateau and the Solar System's largest volcano, Olympus Mons. During this period of heavy bombardment by asteroids and meteors (also experienced by Earth), large impact basins formed, including the Hellas and Argyre Basins. One billion years and some cooling later, Mars had a magnetic field and its atmosphere was dense enough for the planet to sustain liquid water on its surface. Rivers, lakes and possibly even oceans may have formed during this time. Over the next billion years, as Mars cooled further, the planet's dynamo effect ceased and so it lost its magnetic field. Now unprotected, the atmosphere began to thin, stripped of its carbon and oxygen by the solar winds. Its surface water evaporated, leaving behind vast salt flats and dry lakebeds.

Mars's geological activity also slowed due to heat loss and the planet became colder and drier. Wind erosion became the dominant force shaping the planet's surface, creating features such as sand dunes and dust devils (like whirlwinds).

Today, dust storms take place frequently – occasionally covering almost the entire planet for days at a time – and can have severe effects on the global temperature, creating an unstable environment for any life to contend with.

Evidence of water presently on Mars comes from various sources including the Mars Reconnaissance Orbiter (MRO), which has detected seasonal flows on slopes on the planet's surface that are thought to be caused by the presence of briny (salty) water just beneath the surface. They are visible during hotter periods and fade as the planet experiences months of lower temperatures. There is uncertainty surrounding what these flows are, however, as more recent findings and further examination have suggested the streaks may in fact be caused by sand sliding down these slopes instead.

Subsurface water has been located by the Mars Advanced Radar for Subsurface and Ionosphere Sounding (MARSIS) instrument on the ESA's *Mars Express* (2003) spacecraft, which used ground-penetrating radar to detect a large subsurface lake beneath the planet's south pole. This lake is estimated to be about 20 km (13 miles) wide and is thought to be composed of liquid water. NASA's MRO also detected a series of subsurface features that are consistent with the presence of buried ice near the planet's mid-latitudes and believed to be large deposits of frozen water.

Lakes of water below the surface could harbour life, but we need to send technologically far advanced missions to the planet to determine this. The most sophisticated life-searching robot now on Mars is NASA's *Perseverance* (2020) rover (and its companion rotorcraft *Ingenuity*[24]), which is drilling soil samples in Jezero Crater. The ancient impact crater might once have contained floodwaters and today contains clay deposits. *Perseverance* is packaging up some of these samples

---

[24] Nicknamed Ginny, it's a small robotic helicopter.

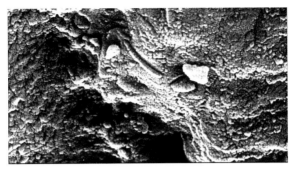

1. The tiny worm-like 'microbe' identified in ALH84001, an SNC meteorite found in Antarctica in 1984. The apparent signs of life in the rock were mostly discredited as the result of non-biological processes. *NASA/JSC/ Stanford University.*

2. Tubeworms and black smokers (deep-sea vents) photographed on an expedition to the Main Endeavour Vent Field along the Juan de Fuca Ridge in the north-eastern Pacific Ocean in 2004. This vent field lies approximately 2,250 m (7,380 ft) below sea level. *Image courtesy of NOAA PMEL Vents Program.*

3. Comet 67P/Churyumov-Gerasimenko photographed by the *Rosetta* spacecraft in 2015. *ESA/Rosetta/NAVCAM – CC BY-SA IGO 3.0.*

4. A tardigrade or 'water bear', one of the most resilient life forms discovered on Earth. *Dr Diane Nelson, courtesy NPS.*

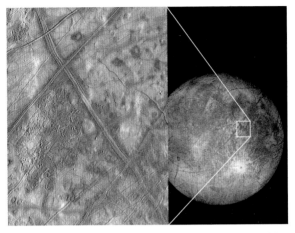

5. The Galilean moon Europa as viewed from the *Galileo* spacecraft. The marks on the moon's surface are the best geologic evidence to date that Europa may once have had a subsurface ocean. *NASA/JPL/University of Arizona.*

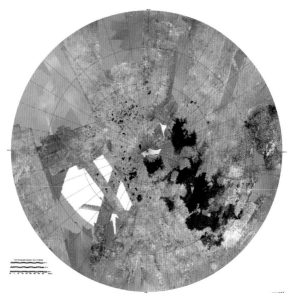

6. A colourised mosaic from NASA's Cassini-Huygens mission showing the lakes and seas on the Saturnian moon Titan's northern hemisphere. Other than Earth, it is the only world in our Solar System known to host stable liquid on its surface. The liquid is mainly methane and ethane. *NASA/JPL-Caltech/ASI/USGS.*

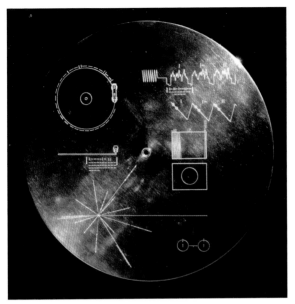

7. The cover of one of the Voyager Golden Records, etched with symbols indicating where the spacecraft that carried it into outer space came from and how to decode the contents. *NASA/JPL-Caltech*.

8. The first deep-field image produced by the James Webb Space Telescope (JWST). The image, captured in 2022, revealed the thousands of galaxies populating just a tiny portion of the Universe. *NASA, ESA, CSA, and STScI.*

in sealed tubes and leaving them on the surface to be collected by a future mission, which may return them to Earth as early as the 2030s.

This rover is able to determine the chemical composition of samples on Mars and search for areas of interest that could have been suitable for life in the past. The state-of-the-art instruments on board have nothing on the capabilities of instruments in labs on Earth. It is very possible, even likely, that the rover's instruments would miss signs of life on Mars if it came across them. To ensure nothing is missed, it is imperative to study these samples back on Earth, where any indications of life can be scrutinised by the wider scientific community and experiments can be repeated for reliability. This is why the most compelling rock samples are being stored for future study in labs that have far more equipment than could ever be sent to Mars on the back of a rover the size of a four-wheel drive.

Ingenuity deserves a mention here not only for being a last-minute test of technology to conduct the first flight on another body in space but also for being incredibly successful, having performed over 50 flights in its first 18 months on Mars. With *Ingenuity* providing a close-up aerial view of Mars, scientists have been able to scout out areas of interest before sending *Perseverance* to see if the region is worth looking at. This saves time and gives a whole new dimension to the mission, as well as paving the way for future flight-based space exploration.

Gas Giants

The four outer planets, Jupiter, Saturn, Uranus and Neptune, are also known as the two gas giants and two ice giants, respectively. They differ greatly from the terrestrial planets, most notably in that they have no solid surface. A lack of a solid surface is a bit of a hinderance when it comes to life; despite Earth being covered mostly in ocean, the water lies on top of the seabed, which is a solid surface. No solid surface means there's nothing for bodies of liquid to sit on and, as we know, a liquid is essential for life's basic processes.

Composed of gases and ices, the planets have cold outer atmospheres. Here, winds regularly reach speeds that would likely tear any living thing to shreds. Jupiter, for example, spins so quickly that winds as fast as 1,400 km/h (870 mph) have been recorded. Beneath the layer of storms in the outer atmosphere of the gas giants, the pressure rises, which causes the temperature to rise as well. We know extraordinarily little about the cores of these planets; scientists believe they may have hot liquid-soup centres or perhaps small, heavy cores like Earth.

The ice giants have a different chemistry to the gas giants. In astronomy, ice does not refer only to water ice – it encompasses the solid form of any condensable molecule that has a freezing point above -173°C (-279°F), essentially anything that can easily change state – from gas to liquid to solid – through variation in temperature. Uranus, for instance, is composed of a water, ammonia and methane slush, which sits around a rocky or icy core. Its atmosphere is mainly hydrogen and helium. Unlike the terrestrial planets, which radiate heat leftover from their formation from their cores, Uranus has no excess central heat so sits at an absolutely freezing -220°C (-364°F). It's far beyond the habitable zone – 19 times further from the Sun than Earth.

It's clear that the surface conditions of most of the planets in our Solar System are not favourable for life. However, not everything needs a solid surface to survive, as evidenced by the marine life on Earth and bacteria and spores that get blown around in the atmosphere. With the clouds of Venus raising some interesting possibilities, the cloud layers of the gas giants might deserve further investigation too.

Moons

While there are only eight planets in our Solar System, there are more than 200 formally recognised moons that come in a variety of shapes and sizes, and with differing environmental conditions. A moon, or natural satellite, is a body in space in orbit around another body that is not a star (also known as a primary). Not only do planets have moons, so do dwarf planets, asteroids and even moons themselves.[25]

Earth's Moon

Our Moon is a great companion to Earth, stopping us from being struck by meteors and giving us tides, which greatly affect marine ecosystems by encouraging the circulation of nutrients and animal and plant life.[26] Lunar rock samples returned by the Apollo astronauts in the 1960s and 70s

---

[25] The moons of moons have been given the ridiculous working title 'moon-moons'...

[26] Plus, it gives us eclipses, which are really cool.

revealed that the Moon is compositionally the same as Earth. In fact, we're quite sure that, billions of years ago, Earth was struck by a Mars-sized planet, which knocked off a piece of our planet that in turn became the Moon. Despite being made of the same stuff, the lunar environment is hugely different to that of Earth. Around a quarter the size of Earth, the Moon has no real atmosphere and its mantle is too cool for tectonic plate activity – it is, therefore, a 'dead' celestial body as it shows no current signs of geological events. Its lack of atmosphere means radiation levels on its surface are high and without geological activity there can be no real change in its habitability for life. Thanks to a rotational period of 27.3 days, one side spends around two weeks at a time facing the Sun, leading to extreme temperature differences between the daytime and nighttime sides (from around -130°C to 120°C (-202 to 248°F)).

We have found no evidence of liquid water on the Moon. It does, however, have PSRs like Mercury, where water ice remains frozen throughout the day and night cycles. This is great news for future human missions that will require water for astronauts and fuel, but the Moon is really quite a hostile place for life and not the focus of any real astrobiology research.

Galilean Moons

Europa

Europa (image 5) is one of the four largest moons of Jupiter and the second of the Galilean moons in terms of distance from the planet.[27] Europa, like our own Moon but a little smaller, is around a quarter the width of Earth. Its atmosphere – barely there – is composed mainly of oxygen but not enough for humans to breathe. Orbiting the Solar System's largest planet, Europa is more than 750 million km (over

---

[27] These moons were named after Galileo Galilei, who discovered them in the early 17th century.

460 million miles) from the Sun. At this distance, very little sunlight and very little heat reach either Jupiter or its satellites. You may think that looking for life on a ball of ice (Europa has an average surface temperature of -171°C (-276°F)) would be a fruitless occupation, but this moon is not solid ice – oh no, it is so much more.

Several missions have been sent to study Europa, including the *Galileo* spacecraft, which orbited Jupiter from 1995 to 2003. The spacecraft provided detailed images of and data on Europa's surface, revealing it to be covered by a layer of ice and having been shaped by a variety of geological processes. The ice is broken by numerous cracks and compacted together into ridges, which suggests that the moon is actively shifting and changing. Some areas of Europa's surface appear to be relatively smooth, an indication that fresh ice may be rising up from below the surface to fill gaps where ice has shifted. This has led scientists to believe that the ice is floating on top of a subsurface water ocean, which may be as deep as 100 km (62 miles). Given its distance from the Sun, you might be wondering how Europa could host any liquid water at all. Well, this is all due to the tidal forces acting between the moon and its primary, Jupiter. Because it takes Europa the same length of time to complete one rotation as it does to complete one orbit around Jupiter (just 3.5 days), the same side of the moon is always facing the planet – Europa is 'tidally locked' to Jupiter.[28] As Europa moves around the planet, Jupiter's gravitational field exerts a force that causes the moon's core to squash and expand, almost as if it were a soft ball being squeezed and released in your hand. This flexing causes the surface ice to crack, forming grooves, and to compact, making ridges where slushy ice from below the surface gets crushed together. All of this flexing causes the moon to heat up, providing warmth for liquid water oceans to exist. Evidence of Europa's oceans

---

[28] Just like our Moon is to Earth!

also comes from data showing Jupiter's magnetic field being disrupted in space by its moon. This discovery led scientists to believe that Europa's weak magnetic field is the result of a global saltwater ocean. Saltwater is conductive (far more so than pure water) and its presence might therefore cause interactions with Jupiter's magnetic field as Europa orbits the planet. Despite this moon's small size, it has been estimated that there is twice as much water in Europa's oceans as in Earth's. With a rocky core, Europa could have vents on its ocean floors (like the hydrothermal vents providing habitats for microbial life on Earth) that outgas warm, nutrient-rich waters and provide an energy source for life. The heat from tidal flexing means warm waters dissolve the minerals from Europa's core, giving rise to an environment that could encourage chemosynthesis.

Studying Europa is challenging due to its distance from Earth and the harsh radiation environment around Jupiter. Its surface is constantly bombarded by charged particles, which can damage spacecraft electronics and instruments.

Ganymede and Callisto

It is thought that two of the other Galilean moons, Ganymede and Callisto, might also host liquid water oceans, although the evidence for this is not as strong as that for Europa. Ganymede is the largest moon in the Solar System and it has a magnetic field believed to be generated by a partially molten iron core. Its surface is covered in craters, ridges and grooves, and also features a vast network of interconnected fractures, resembling the rift valleys that form when tectonic plates move apart on Earth. Callisto is Jupiter's second largest moon (and third largest overall) and it has the most heavily cratered surface in the Solar System. This suggests it has not experienced significant geological activity in billions of years, as smoother surfaces are associated with more active worlds. When geological activity is high, lava flows fill in the craters left by the impacts of asteroids or comets and cool, smoothing a planet or satellite's surface once more. Both moons are thought to have subsurface oceans but are yet to be explored in detail. Future missions should give us an insight into whether they could be home to life.

Saturn's Moons

Enceladus

Enceladus is one of the more than 80 known moons of Saturn, discovered in 1789 by William Herschel. It is relatively small, with a diameter of about 504 km (313 miles), and it orbits Saturn at a distance of about 1.5 billion km (930 million miles) from the Sun. It should come as no surprise that it too is an icy moon, but it is covered in new, clean ice, indicative of geological features such as **cryovolcanoes** that regularly furnish its surface with fresh ice. Instead of firing out lava, dust and ash, cryovolcanoes spew out ices, liquids and vapours. Enceladus is also notable for its geysers, which shoot plumes of water vapour and ice particles from its surface into space. For its size, these plumes are mightily impressive. They expel material hundreds of kilometres above the moon's surface at a rate of 400 m/s (1,300 ft/s) and are responsible for one of the rings of Saturn – its E ring. They were first discovered by the *Cassini* spacecraft during multiple flybys of Enceladus in 2005 and subsequent observations have revealed that they contain organic molecules, indicating that Enceladus may have the ingredients necessary for life. Many different hydrocarbons (organic molecules made up of only hydrogen and carbon, such as methane) were detected in the plumes, as well as everything needed to form amino acids, including phosphorus, nitrogen and oxygen. The moon's surface is also marked by a variety of features, among which are impact craters, fractures and ridges. One of the most striking features are the 'tiger stripes', a set of four parallel fractures near its south pole, that are the source of the geysers. These fractures are thought to have been created by tidal forces exerted by Saturn, which have the same effect as Jupiter's tidal forces on Europa and cause its interior to heat and flex. A possible subsurface ocean of liquid water may interact with the moon's rocky core and create hydrothermal vents. This, along with the organic molecules detected in the plumes, has made Enceladus a prime target of astrobiology research.

Titan

Titan, Saturn's largest moon, is bigger than the planet Mercury. It's the only moon in our Solar System with a significant atmosphere. Titan's atmosphere, like Earth's, is mostly composed of nitrogen, but with a thick layer of organic compounds, lending it a hazy, orange-brown appearance. This atmosphere is opaque from the outside, obscuring visibility of the moon's surface, so it was not until the NASA Cassini-Huygens mission (1997–2017) that we were able to see what lay beneath the dense clouds – it did not disappoint. In January 2005, after the mission reached Titan, the *Huygens* probe was released from orbit and fell to the surface of the moon. Before coming to a planned end on impact with the ground, it had transmitted invaluable data. The images captured revealed a landscape that resembled Earth, with rivers, deltas and even waves – yes, waves, meaning there was liquid on its surface (image 6). Following water has been key to investigating habitable conditions on other planets, but we know that other liquids may be of interest as alternative solvents for life too.

Titan's liquid lakes and seas are composed not of water but of hydrocarbons, including methane and ethane. Usually found as gases on Earth, these substances are liquid on Titan due to its low surface temperature of around -179°C (-290°F). Titan is the only place in the Solar System other than Earth that has been found to have stable bodies of liquid on its surface; it even has an entire methane cycle that operates in the same way as the water cycle on Earth. Liquid methane on its surface evaporates into clouds, condenses as it cools and then falls from the clouds back to the surface as methane rain. It is thought that life different to that on Earth could use liquid methane in the way the life we know uses liquid water. The surface of Titan is covered with a thick layer of organic materials, including complex molecules that could be the precursors to life. Among those identified are nitriles, formed of a nitrogen-containing hydrocarbon, and benzenes (which are produced by natural processes, like volcanic activity, on Earth).

In addition to its geological features, Titan also has a magnetic field thought to be generated by an internal dynamo, caused by the interaction between the moon's core and its atmosphere. The presence of both a thick atmosphere and a magnetic field are positive signs for any potential life, as both offer protection from harmful radiation.

Titan's atmosphere provides enough pressure that humans would not require a protective pressure suit to walk on its surface, but they would need an oxygen mask and insulation from the freezing temperatures. Due to Titan's size, it exerts a gravitational force similar to the one exerted by our Moon (about 1/6 Earth's gravity), meaning that raindrops on Titan would fall much slower than they do here (travelling at 1.6 m/s² (5.25 ft/s²) or 1/6 of the speed) and they are also around 50% larger. There is something quite magical about the thought of slow-falling, big raindrops, although preferably somewhere with a warmer climate!

If Titan's surface liquid wasn't exciting enough, like the other moons mentioned so far, it is also thought to have a subsurface ocean of liquid water (and not one composed of methane!). As a geologically active moon, it could also be home to potentially life-sustaining hydrothermal vents.

Dwarf Planets

In 2006, the ever-popular Pluto was demoted to dwarf-planet status by the International Astronomical Union (IAU), much to the distress of many Plutophiles. It is not alone in its classification though; there are five officially recognised dwarf planets in the Solar System and we expect this number to grow in future.

Proximity to the Sun may not matter if an astronomical body can gain heat from another source, so the environments on some dwarf planets may be (or may once have been) conducive to life. Dear old Pluto is nestled in the Kuiper Belt, around 6 billion km (3.7 billion miles) from the Sun, so was thought to be an icy, cold rock, but data from NASA's 2015 *New Horizons* probe hinted that it may not be so lifeless after all. Its icy outer shell is thought to conceal subsurface liquid oceans that may have formed early in its history. As with the planets, before it cooled to its present state, Pluto would have been hot, so could have had liquid water that may remain beneath the surface, offering a home to potential life. Pluto and its largest moon, Charon, both have tholins on their surfaces, which is a term for the gunk-like organic molecules that appear in images as orange-red patches. We know little else about them or their origins.

The ESA's infrared Herschel Space Observatory recorded Ceres, a dwarf planet found in the asteroid belt, spewing water vapour from its icy surface when being heated by the Sun. Water evidence from objects in the Kuiper and asteroid belts leave open the possibility that asteroids and comets like Themis and 67P could also harbour some of the necessities for life.

'Oumuamua

'Oumuamua, the 'messenger from afar', was caught hurtling through the Solar System on 19 October 2017 by astronomers using the Pan-STARRS1 telescope in Hawai'i. The comet or asteroid caught the attention of scientists due to its unusual characteristics and path. Highly elongated and resembling a cigar or needle with an estimated length of about 400 m (1,300 ft), the visitor's distinct shape was notable compared to other similar objects observed in our Solar System. Its high velocity, up to 87.3 km/s (54.2 miles/s) at its closest approach to the Sun, and odd trajectory suggested interstellar origins. Although Solar System comets tend to have highly elliptical (oval) orbits, 'Oumuamua's path was far more exaggerated than we usually see. Scientists proposed several hypotheses: it could be a fragment of a larger object that was ejected from another star system or a remnant of a planet that did not fully form. There were even suggestions that it could be a fully operational interstellar spacecraft. This final idea was thrown around at the time but there is no evidence at all to support it, however exciting that would be! Despite just an 11-day window for observation, 'Oumuamua sparked significant scientific interest. Here was a unique opportunity to study an interstellar object, a reminder that life may not be limited just to our Solar System.

Other Solar Systems and Exoplanets

Exoplanet detection

Over the past 30 years, we've been searching for planets that orbit other stars. These are called extrasolar planets, or exoplanets for short. So far, we've discovered over 5,000 and there are at least another 5,000 potential candidates that need further observation until they can be confirmed as such. The data from our short time exoplanet-hunting has indicated that most stars host planets and for every star there appears to be, on average, at least one planet. In the Milky Way alone there are somewhere between 200 and 400 billion stars. By this measure, we could expect there to be at least as many exoplanets, with some stars hosting many planets and others none at all. Of those exoplanets that have been discovered, only around 4% are defined as terrestrial (between half Earth's size and twice its radius), while 31% are so-called '**super-Earths**' (more massive than Earth but not necessarily similar in other ways – they may have large atmospheres and potentially no solid surface). There are also Neptune-like exoplanets, including mini-Neptunes, and gas giants, some of which are classed as 'hot Jupiters'.

The greatest challenge we face searching for life on these distant worlds is, well, the distance. The closest star to our own star is Proxima Centauri, which is host to our closest exoplanet, Proxima Centauri b – just 4.2 light years away, a mere 40 trillion km (25 trillion miles). With current technologies, that journey would take a staggering 70,000 years. Sending humans to check out these potentially habitable planets is not practical and may never really be possible. Instead, it is up to our scientific instruments on Earth or, better yet, in space, to gather more information for us.

There are many ways to detect exoplanets, but the method that has so far yielded the best results is the **transit method**. This involves looking at a star and monitoring its light intensity over a period of time (Figure 5). A regular dip in that intensity suggests there could be a planet orbiting the star. As a planet passes in front of a star, it blocks out some of the light emitted by that star. Once the planet is no longer in front of the star, the light intensity we detect will return to normal. We can witness transit events in our own Solar System. The most common are eclipses, like solar eclipses, when the Moon passes in front of the Sun and blocks a portion (or sometimes all) of its light from reaching Earth. There are also planetary

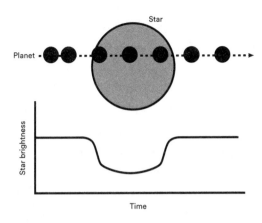

Figure 5. The transit method – as a planet passes in front of a star it blocks some of the star's light, causing a regular dip in the intensity of light emitted.

transit events and we can observe two from Earth: the transits of Mercury and Venus, the planets closer to the Sun than Earth. During these transits, we can see the circular shape of the planet passing in front of the Sun. They happen far less often than eclipses; the most recent planetary transit was that of Mercury in 2019. The next Mercury transit won't take place until 13 November 2032 and the next transit of Venus is not until 2117. They are, however, vital for our understanding of transit events further afield.

As part of the transit method, scientists measure the amount of infrared light (which humans cannot see) being emitted by a star and form a light curve. Its intensity changes more noticeably than that of visible light during a transit event. The curve is used to determine a possible exoplanet's various properties, including the length of its orbit, its size and the speed at which it travels around its host star.

The Kepler Space Telescope was a NASA mission designed to discover exoplanets. Operating between 2009 and 2018, it made significant contributions to our understanding of the Universe. Kepler used the transit method to monitor continuously the brightness of stars in a specific region of the Milky Way and it discovered over 2,600 exoplanets. Using this data, scientists have determined that small, terrestrial planets are common around low-mass stars (like our Sun) and that up to half of all such stars in the Milky Way may have terrestrial planets in their habitable zone.

Exoplanet Kepler-62f is a super-Earth that lies within its star's habitable zone and was discovered by the planet hunter in 2013. Based on its distance from its star and its size (around 1.4 times the radius of Earth), it has been deemed a promising candidate for habitability. There are, however, still many unknown aspects of Kepler-62f's environment that we will need to explore to determine its habitability, especially its atmosphere.

Atmosphere characterisation

New space telescope missions will help us characterise the atmospheres of exoplanets, revealing vital clues as to their overall composition. Scientists will be looking out for molecules of water vapour, in particular. As we all know by now, water is very important!

The James Webb Space Telescope (JWST) was launched successfully on Christmas Day 2021 and has since sent back incredible images, as well as data on exoplanet atmospheres. Using the same principles of spectroscopy employed when looking for biosignatures, the telescope is revealing that the atmospheric conditions of giant planets do not fit the pattern of gas giants orbiting our Sun. JWST tracks planets as they pass in front of their stars. As this happens, some of the star's light filters through the planet's atmosphere. By comparing the signature spectrum of the star and the combined spectrum of the star and planet, researchers can determine which molecules are present in the planet's atmosphere. Through the use of infrared light, the telescope can also detect chemical fingerprints, which cannot be detected in visible light. The atmosphere of exoplanet HD 149026 b (known as Smertrios) in the constellation Hercules, shows an overabundance of carbon and oxygen, even though it has a mass similar to Jupiter. The giant planets in our Solar System have much lighter hydrogen and helium atmospheres so it is unusual to observe these heavier elements in the atmosphere of an exoplanet of this mass.

Meanwhile, WASP-39 b, a gas giant exoplanet orbiting a star 700 light years away, provided JWST with its first detection of sulphur dioxide, a molecule produced through chemical reactions induced by the emission of high-energy light from the parent star. This discovery suggests the **photochemistry** process (relating to the absorption of UV, visible light or infrared radiation) plays a crucial role in the atmospheric composition of exoplanets. The presence of sulphur dioxide in the atmosphere of WASP-39 b is unusual; volcanism on terrestrial planets such as Earth and Venus, as well as on Jupiter's moon Io, releases sulphur dioxide into the atmosphere, but on gas giants like Jupiter and WASP-39 b without this type of geological activity it must come from a different source.

The TRAPPIST-1 star system is the most studied planetary system after our own and lies just 40 light years away. First discovered in 1999, this dwarf star in the constellation of Aquarius was later found to host not one but, as of 2017, seven planets. At 2,300°C (4,172°F), compared to our Sun it's relatively cool. The exoplanets (named TRAPPIST-1b to h, with b being closest to the host star and h furthest) are of particular interest. They are all around the same size as Earth (between 0.76 and 1.13 times its radius) and are thought to be of a similar composition (mostly rocky). Planets e, f and g lie within the star's habitable zone and the furthest planet, TRAPPIST-1h, orbits six times closer to the star than Mercury does to the Sun. The whole system is extremely compact. This leads to the possibility that the planets are exerting forces on each other, tidal flexing, as they complete their orbits. In fact, the planets would appear in each other's skies, just as the Moon does in our sky. JWST's first release of data also contained thermal signatures of TRAPPIST-1b, indicating that it apparently has no atmosphere and would therefore not be habitable.

Observations conducted with JWST's Near-Infrared Spectrograph (NIRSpec) may also have found evidence of water molecules in the atmosphere of the rocky exoplanet GJ 486 b. What's intriguing are the extreme conditions these molecules may be present under: the planet is extremely close to its star, far too close to be within its habitable zone, resulting in a scorching surface temperature of approximately 430°C (800°F).

There is a caveat, though: the water vapour may not originate from the planet itself but could be linked to starspots – like sunspots, these are darker, cooler areas on the surface of a star. GJ 486 b orbits a red dwarf star at close proximity. It is most likely tidally locked to its host star, with permanent day and night sides. For it to maintain an atmosphere with water vapour under such extreme conditions, a geological process like volcanic activity may be necessary to continually replenish its surface with liquid. JWST's other instruments will conduct further observations of GJ 486 b to help unlock its mysteries and see if it really could be a habitable world.

The telescope's revelations are exciting scientists, who expect JWST to contribute significantly to the understanding of exoplanet atmospheres. Its infrared light analysis offers an entirely new view of exoplanet atmospheres – we are seeing them, quite literally, in a new light.

Changing habitability

Just as animals go through life cycles, so do stars. As hydrogen atoms are combined to form helium in a process called nuclear fusion, stars 'burn' their fuel.[29] The habitable zone around a star, where conditions are ideal for liquid water to exist, will shift over time as the star continues through its life cycle.

Let's take our own Sun, a main sequence yellow dwarf star, as an example. It's fairly average and only of particular interest to us as our closest star. In about 4 billion years, when the Sun has used up all its hydrogen, it will begin to expand, pushing away its outer layers of gas. They will, in turn, become more diffuse. Eventually, the layers of gas will engulf Mercury, Venus, Earth and possibly Mars too. Ignoring the fate of Earth(!), this stellar expansion will bring heat to those bodies in the outer Solar System that have escaped consumption. It will warm places that are currently thought to be too cold for life and it may well melt the thick, icy surfaces of the moons of the gas giants.

---

[29] Burn is perhaps one of the worst terms that could be used to describe this process, as it has led to the common misconception that there is fire in the Sun and other stars.

The exact consequence of this future heating with regards to creating habitable environments is impossible to predict, due to the number of parameters involved and many potential interactions of innumerable factors. The point is to keep an open mind when considering if a place could sustain life. It might appear utterly inhospitable at present but could well have been a very different place in the past and may yet transform again in the future. The entirety of a solar system – not a singular planet or astronomical body – should be taken into account, with due consideration for its evolution and any future events that could affect habitability.

Search for Intelligent Life

Most current research on the search for life is looking for signs of microbial life, but we are also looking for complex life, even intelligent life. We're fairly certain that there are no walking or talking aliens in our Solar System unless they're doing a really good job of hiding. So, we need to look further afield for intelligent life – to other solar systems and even other galaxies.

Earlier, we discussed biosignatures used as evidence of the presence of current or past life. Some are barely observable with the naked eye, others more noticeable – a molehill in the park, for example.

Intelligent life may show far larger signs of interaction with their environment.

In 1985, a hole was discovered in Earth's ozone layer, part of the atmosphere that protects us from damaging solar radiation. It was determined that chlorofluorocarbons, better known as CFCs and found in aerosols, solvents and refrigerators, were responsible. Their production was soon banned worldwide. This disturbance to the atmosphere could be detected from space and it would indicate to an outside observer that something not entirely natural, or something highly unusual, was happening on Earth. It's possible that we could detect something similar on another planet if an alien civilisation also had the technology to vastly alter its atmosphere.

In looking for signs of more technologically advanced civilisations, we might be looking for structures we have theorised but are not currently able to construct ourselves. One such hypothetical megastructure is the Dyson sphere. It was popularised by the physicist and mathematician Freeman Dyson (1923–2020), who was inspired by *Star Maker*, a 1937 work of science fiction by Olaf Stapledon (1886–1950). The Dyson sphere was a thought experiment, proposed as a solution to the ever-increasing energy demands of an advanced spacefaring civilisation that needed more energy than it could get from its home planet's position relative to its star. The hypothetical civilisation would be required to build a structure around the star that was capable of harnessing far greater amounts of its energy. The idea has since been extended to constructing similar things around pulsars (fast-rotating stars that send out strong pulses of EM radiation at regular intervals) and even black holes. In theory, a structure such as this could be detected using spectroscopy, as its emission spectrum would be obviously different to that of a star or other astronomical object without a similar structure around it.

Organisations like the Search for Extraterrestrial Intelligence (SETI) Institute are actively scanning areas of the sky to listen out for abnormal radiation, such as radio signals. There are many objects and events in space that give off radio signals, including black holes. It's not just a case of listening for any radio signal, but identifying unusual signals.

If there are advanced aliens that can build technology, we would expect to be able to detect them using radio telescopes. This might seem a tad presumptuous, but bear with me. Considerable amounts of our advanced technology give off radio waves, including radios (hence the name!) and televisions. So, when humans turned on the first television, it started sending out its signal into space. The very first radio message was transmitted in 1900. By now, those human-produced radio waves will have travelled about 120 light years into space.

Radio waves can travel very long distances and can easily pass through atmospheres, unlike some other types of light. That's why they are the wavelength of choice for long-distance communication – we use them on spacecraft to send messages from Earth into space and back. This property would make them ideal for any intelligent alien civilisation to use in communication both on its home planet and perhaps to send messages to us across the cosmos. How can we be sure this is the correct method? Well, truth be told, we can't. Observations and experimentation tell us the laws of physics are universal. Thus, if radio communications are the best way for us to communicate from here on Earth, it is likely to be the same for any other intelligent life, wherever it may reside in the Universe.

'Wow! signal'

Back in the 1950s, scientists thought they might indeed have detected an unusual signal. Listening out for a signal does not mean donning a pair of headphones in the hope of hearing the far side of the Galaxy's Top 40 chart hits. Data is collected and, in the past, it would have been printed out by machine. This could then be reviewed by astronomers, who could recognise anything unexpected in the printout. In August 1977, American astronomer Jerry R. Ehman did just that when he noticed an unusual reading on the printout he was reviewing from Ohio State University's Big Ear radio telescope. He duly circled it and wrote the word 'Wow!' next to the anomaly, giving rise to the name the 'Wow! signal'. Astronomers began looking for explanations, contacting other observatories to see if they had picked up something similar around that time. To this day, the origin of the signal has not been identified, although it is unlikely to be from a source of extraterrestrial communication as it was very short and has not been observed again since. If ET was trying to phone home and his call went unanswered, you would assume he would at least try once more. Still, the signal is the kind of transmission from extraterrestrial intelligent sources that scientists are hoping to find.

We could be completely wrong, though. There may be civilisations far more advanced than us, which communicate using technology we have yet to discover. Alien communications that we are too primitive to understand may be bombarding Earth daily. Even if we do recognise something as a form of alien communication, we may never be able to interpret it. Of all the species on Earth, we can only communicate with a very small portion of them on an incredibly basic level. Aliens may have the same difficulties communicating with us.

The Drake equation and the Fermi paradox

The Drake equation was formulated in 1961 by Frank Drake (1930–2022), a former director at the SETI Institute, and it uses a list of parameters to try and estimate how many advanced civilisations that we could communicate with exist in our Galaxy (Figure 6). This definitely wasn't an exact science and the variables are very difficult to quantify, but it was an attempt to understand broadly what we're looking for, how to find it and where. The original values used in the equation resulted in a number (N) of between just 20 and up to 50 million. That seems quite promising, I'd say. But science has changed a lot since the 1960s. We discovered the first exoplanet in the 90s and have found thousands more since. As a result, we have a smaller range of potential values for each of the components of the Drake equation. New estimates give answers between 0 and 15 million. We can arrive at 0 thanks to the 'Rare Earth hypothesis', which states that such a large combination of things was required to give rise to life on Earth that it was in itself an incredibly unlikely outcome. The same combination of events and conditions couldn't possibly happen elsewhere again.

$$N = R \times f_p \times n_e \times f_l \times f_i \times f_c \times L$$

N = The number of civilisations in our Galaxy with which communication might be possible.

R = The average rate of star formation in our Galaxy.

f_p = The fraction of those stars that have planets.

n_e = The average number of planets that can potentially support life per star that has planets.

f_l = The fraction of planets that could support life and actually develop life at some point.

f_i = The fraction of planets with life that go on to develop intelligent life (civilisations).

f_c = The fraction of civilisations that develop a technology which releases detectable signs of their existence into space.

L = The length of time for which such civilisations release detectable signals into space.

Figure 6. The Drake equation is used to estimate the number of advanced civilisations in existence in our Galaxy with which we could communicate.

This is an anthropocentric view, which assumes that other life must be like life on Earth, when we really don't know this to be the case. We only have one example, which any scientist will tell you is not a reliable sample size!

But, if the Drake equation suggests that there could be up to 15 million fully technological civilisations in our Galaxy... then where are they all? Shouldn't we be hearing them, picking up the signals from their television shows?

This idea is called the Fermi paradox and many solutions to it have been put forward. One states that maybe we're the first – that we evolved very quickly given Earth's 4.54 billion years. Another proposes that there were other civilisations, but they were wiped out by a meteor strike (similar to the one that caused the extinction of the dinosaurs) or some other cataclysmic event. There are numerous potential explanations for why we have not found life when the numbers suggest otherwise. Just because we have not yet found something, does not mean it is not there.

Frank Drake and Carl Sagan were true visionaries when it came to the search for extraterrestrial intelligence and came up with many ideas to further it. The Arecibo message was a binary-coded message (consisting of ones and zeroes) that was beamed into space on 16 November 1974 from the Arecibo radio telescope in Puerto Rico. The message was aimed at the centre of a globular star cluster known as M13, in the constellation Hercules, which is located approximately 25,000 light years from Earth. M13 was the chosen target simply because it was in the right place in the sky for the ceremony and not too far away – in astronomical terms, at least! The emission was an omnidirectional transmission, meaning that it was designed to be picked up by a telescope similar in size to Arecibo just about anywhere in our Galaxy. A demonstration of technological achievement, it was also intended to provide extraterrestrial beings with information about the human race and our understanding of the Universe. The message contained information about the structure of DNA, the human population, the Solar System and other details about Earth and humanity. As the signal travels at the speed of light, it will take 25,000 years to reach its destination, by which time the centre of the cluster will have moved. A point worth noting when trying to communicate across large distances: by the time your signal gets to the area it was directed towards, your intended target may have moved.

Pioneering missions

With its Pioneer missions, NASA sent a series of unmanned space probes to study the outer Solar System. *Pioneer 10* and *Pioneer 11* were launched in the early 1970s and both carried a plaque intended to convey a message to any extraterrestrial life they might encounter. The plaques were designed by, you guessed it, Carl Sagan and Frank Drake, and made of durable gold-anodised aluminium. They featured a drawing of a man and woman holding hands as a sign of unity, along with several other symbols that were intended to share information about the Solar System and our place in the Universe. These included a diagram, showing the positions of the planets relative to the Sun, as well as a representation of the hyperfine transition of hydrogen, which is a universal physical constant that can be used to determine the units of time and distance used on the plaque.

The Voyager missions were launched two weeks apart by NASA in 1977 to study the outer Solar System and beyond. Like the *Pioneer* probes, the two *Voyager* spacecraft were equipped with more than just scientific instruments – each included a 'Golden Record' (image 7). The records were compiled by a team led by our favourite astronomer Carl Sagan[30] and they contained a variety of sounds – including greetings in 55 languages, music, laughter and animal noises – and images, selected to represent the diversity of life and culture on Earth. Among the photographs were pictures of landscapes and human anatomy, as well as diagrams explaining scientific concepts.

The Golden Records were placed on the *Voyager* spacecraft because they were expected to travel beyond our Solar System

---

[30] Well, he's mine, at least.

and potentially encounter extraterrestrial life. While the likelihood of the records being found and understood by any intelligent alien beings is low, the Voyager missions remain an important milestone in our efforts to explore the Universe. Both *Voyager* spacecraft have now left the influence of the Sun (the heliosphere): *Voyager 1* in 2012, followed six years later by *Voyager 2* in 2018. They're now further from Earth than any other human-made objects have ever been. As of July 2023, *Voyager 1* is nearly 24 billion km (15 billion miles) from Earth, meaning it takes over 22 hours for a signal to reach it from Earth.[31] The immense distances these spacecraft have travelled cements our legacy as interstellar explorers.

One day, we may be looking at another planet and we'll see something that just seems out of place. It's possible that the most compelling biosignature will be one that hasn't yet been identified

---

[31] 44 hours to send a signal and receive one back!

as a potential sign of life. This is where human observation and study really stand out compared to robotic or telescopic exploration. Humans have an instinct to acknowledge if something is unusual; we may not know the cause, but we can tell when it deserves further investigation. This may be our best bet when looking for life elsewhere. In truth, there is no way to know that life elsewhere will develop and behave in the same way as we have come to expect life on Earth will. It may be that the only indication of other life is a realisation that there is something going on that we do not understand or cannot account for by any other reasonable explanation.

Future Missions and Possibilities

Recent research into phosphine signals detected in Venus's atmosphere has reignited interest in the planet. Six missions are under development, including an orbiter and atmospheric probe from NASA, both of which will investigate the cause of this unusual detection. Many more space exploration missions have been proposed. Some, like the Japanese *Akatsuki* orbiter, aim to deepen our understanding of Venus's atmosphere, surface and geological processes, while others lack a goal specifically linked to astrobiology but will help expand our knowledge on the topic nonetheless. Space agencies from countries worldwide are putting their efforts into this new age of space exploration – it's an incredibly exciting time.

As part of the ESA's ExoMars mission, a rover named *Rosalind Franklin*, equipped with a 2 m-long (6 ft 7 in.) drill, is hoping to head off on its journey to Mars in 2028.[32] After landing on the Red Planet, the rover will drill into its surface and use its onboard laboratory to search for biomolecules and other biosignatures. Mars's lack of atmosphere leads scientists to believe that if there any signs of past or present life on Mars they would be underground, shielded from the harsh surface conditions. The rover's drill will open up this area for investigation by astrobiologists for the first time.

Plans are underway to send humans to Mars, which will enable us to study the planet first-hand. Private company

---

[32] Its departure has been delayed since 2016.

SpaceX has been awarded a contract by NASA to deliver a crew vehicle as part of its Artemis programme, which aims to return humans to the Moon. SpaceX intends to use a modified version of the *Starship* it has been working on for this mission. The vehicle's main purpose, however, is to carry crew and cargo to Mars as part of the company's goal to establish a permanent human presence on the planet. The first crewed mission to another planet will be a hefty undertaking and the current timelines seem incredibly ambitious.[33] However, it is difficult to quantify the benefit to planetary science that human missions could bring. Considering the contributions to lunar science made by the Apollo missions to the Moon, there can be no doubt that an equivalent mission to Mars would prove invaluable. To put it into perspective, over the almost three-and-a-half days those 12 astronauts collectively spent outside the

---

[33] NASA has plans for manned missions as early as the 2030s.

lunar module on the surface of the Moon, they covered a distance of 95 km (60 miles). NASA's *Opportunity* rover, on the other hand, was operational on Mars from 2004 to 2018 but only traversed 45 km (27 miles). The later Apollo missions were assisted by a lunar rover, so the distance was not all covered on foot, but the difference between solely robotic and human missions is staggering. Having a human presence on Mars would revolutionise our understanding of the planet and planetary science more widely. Being able to take a closer look at something that just doesn't seem right could lead to the discovery of life in a way that would be impossible to achieve solely with a robot or lander.

Icy moons

Soon, we'll be able to review data from missions to the moons of the outer planets too. The ESA's *Jupiter Icy Moons Explorer* (JUICE) and NASA's *Europa Clipper* are set to arrive at Jupiter in the early 2030s. Both are orbiters. *JUICE* will focus on Ganymede, Europa and Callisto, while NASA's mission intends to conduct 45 flybys of Europa. The benefit of these missions overlapping is that scientists will be able to cross-reference data if either one detects something of interest. Conceptual plans for landers on these icy bodies have also been proposed. One includes an ambition to drill deep into the surface of Europa and deploy what is essentially an autonomous submarine into the oceans below. We'll have to wait and see if these grand plans come to fruition.

NASA's *Dragonfly* spacecraft is due to launch in 2027. Headed for Titan and with a scheduled arrival date of 2036, it's set to study the Saturnian moon's prebiotic chemistry. Following the success of the *Ingenuity* helicopter on Mars, this rotorcraft will conduct a series of flights across Titan, covering up to 8 km (4 miles) each time. Then there's Breakthrough

Enceladus, a collaborative project between NASA and a private company that will explore the possibility of microbial life in the warm water jets of another of Saturn's moons.

When it comes to what space telescopes can show us, we've only scratched the surface. JWST has only recently begun its mission. As the successor to Hubble, which itself has given us 30 years of unbelievable images and data, we can expect even more tantalising results to come. Alongside observing exoplanets and providing more information about their atmospheres, it will look further into the Universe than any telescope has done before. It has already produced its own deep-field image, following on from the Hubble Deep Field and Hubble Ultra-Deep Field images that changed our view of the Universe forever. These photos were taken of a small patch of seemingly empty sky, covering an area equivalent to that covered by a grain of rice if it were held at arm's length. The images, each with an exposure time of several days, revealed an area teeming with galaxies. Since then, deep-fields have been taken of multiple areas of what we once thought was empty space and produced the same results. With its first such image, JWST revealed the oldest known galaxies and will only further expand our knowledge of the Universe (image 8).

The ESA will launch its Planetary Transits and Oscillations of Stars (PLATO) telescope in 2026, the aim of which is to determine the characteristics of rocky exoplanets. It will hunt for Earth-like planets around Sun-like stars and seek other planets orbiting the habitable zone of their parent star. The telescope will also analyse the stars to gather data on starquakes – seismologic activity on stars, like earthquakes on Earth – and learn more about them.

Launching in the mid-2020s, NASA's Nancy Grace Roman Space Telescope will seek over 2,000 exoplanets in the inner Milky Way using a wide-field infrared camera with a field of view 100 times larger than Hubble's!

Solar sails

The extreme travel times to reach even our closest star system are inconceivable on a human scale. There are missions in the works for an interstellar trip using solar sails, though. A solar sail is an extremely thin and light piece of metallic material that can be propelled using a laser. As photons hit the sail, the craft is pushed in the opposite direction. With continuous contact, the sail's speed will gradually increase to up to 0.05% the speed of light (about 15,000 km/s or 9,300 miles/s). The weight constraints mean that there is no prospect of this technology being of use in human space exploration, but it could be used to send very small instruments extremely long distances, giving us an inside view of another solar system. Any data that does get sent back from a star like Proxima

Centauri would still take 4 years to reach Earth, as we are limited by the finite speed of light.

Will we ever find other life?

Our knowledge of the Universe we inhabit and the science it is governed by is always increasing, whether that be through the confirmation of a hypothesis long believed to be correct or proof that another claim is not. Searching for life is an endeavour that will never cease. If life is discovered somewhere as close to home as the subsurface of Mars, scientists will only be spurred on to search for other life in more distant locations. It seems almost certain there must be life somewhere else in the Universe due to the sheer numbers involved, but it is so vast that extraterrestrial life may just be out of reach forever. We've discovered a huge number of exoplanets and countless moons, dwarf planets, comets and asteroids – many more remain to be found. You don't need to be a statistician to see that it's incredibly unlikely that the only life in the Universe is the life we see on Earth, given the number of potentially habitable environments in the Milky Way alone.

There is the further possibility that the Universe is infinite. If that's the case, there would, of course, be infinite planets. Wouldn't that mean there's an infinite probability life would exist elsewhere too? Well, it might, but the trouble is that it could be so far away that we would never find it. The search for life may be never-ending, but it's a journey filled with excitement, wonder, awe and knowledge. It expands our horizons and encourages us to learn about ourselves along the way. If we really are all there is in the Universe, then all life on Earth is unfathomably rare. Should we not have an unwavering respect for it? And, furthermore, if there is other life, we should consider how we, as one species, as one planet, want to present ourselves to the rest of the Universe.

I encourage you to look up to the night sky and wonder what it means to be human, to be a part of the Universe, whether we are alone or part of an entire unknown cosmic community and what that means to you.

'We are either alone in the Universe or we are not. Both possibilities are equally terrifying.'
Arthur C. Clarke (1917–2008)

Glossary

abiogenesis – the hypothetical process by which life arises from non-living matter, such as simple organic compounds; also known as spontaneous generation.

(ocean) acidification – a reduction in the pH of the ocean over an extended period, caused by an increase in the amount of carbon dioxide absorbed from the atmosphere.

anabolism – a metabolic process that involves the making of complex molecules from simpler ones.

astrobiology – the study of life in the Universe, including its origins, evolution, distribution and future.

catabolism – a metabolic process in which complex molecules are broken down into simpler ones.

chemosynthesis – the process by which living organisms convert chemicals into food, often in the absence of sunlight as a source of energy.

cryovolcanoes – 'ice volcanoes' that erupt ice, liquids and vapours (like water, ammonia and methane) instead of magma.

cosmic pluralism – a belief in the existence of numerous worlds in addition to Earth that may harbour extraterrestrial life.

extremophile – an organism capable of thriving in extreme environments, specifically those that test the limits of what known life is able to adapt to.

genome – the entire set of DNA instructions found in a cell.

habitable zone – also known as the Goldilocks zone, the region around a star where liquid water can exist on the surface of surrounding planets.

horizontal gene transfer – the movement of genetic material (DNA or RNA) between unrelated organisms without reproduction.

metabolism – the life-sustaining biochemical reactions within an organism.

organic molecule – a molecule typically found in living organisms containing either a carbon-carbon bond or a carbon-hydrogen bond.

polymer – a substance consisting of large molecules made up of repeating patterns of smaller molecules.

photochemistry – a chemical reaction relating to the absorption of ultraviolet (UV), visible light and infrared radiation.

photon – a particle of light.

prokaryote – a single-celled organism without a distinct nucleus or any other subcellular structures encased in a membrane. Bacteria are prokaryotic organisms.

panspermia – a hypothesis that suggests that life exists throughout the Universe and may be transferred between planets and through space via bacterial spores in dust or on asteroids and comets.

super-Earth – a class of planets more massive than Earth but lighter than the ice giants (between twice the size of Earth and up to 10 times its mass). May be composed of gas, rock or a combination of both. No super-Earths exist in our Solar System, but they are common in our Galaxy.

spectroscopy – the study of the absorption and emission of electromagnetic radiation by matter, using spectra.

transit method – a method of detecting planets outside our Solar System that measures the intensity of light being emitted by distant stars for periodic dips in brightness. A regular dip would indicate that the star's light is being blocked from reaching the observer by another astronomical body (such as a planet) passing in front of the star as it orbits.